THE
PROFESSIONAL ENGINEERING
CAREER DEVELOPMENT SERIES

CONSULTING EDITOR

Mr. Dean E. Griffith, Director, Continuing Engineering Studies, The University of Texas at Austin

CONSULTANTS

Dr. John J. McKetta, Jr., E.P. Schoch Professor of Chemical Engineering, The University of Texas
Dr. Maurits Dekker

EDITORIAL ADVISORY COMMITTEE

Dr. Walter O. Carlson, Acting Dean of Engineering College, Georgia Institute of Technology
Mr. R.E. Carroll, Director, Continuing Engineering Education, The University of Michigan
Mr. William W. Ellis, Director, Post College Professional Education, Carnegie-Mellon University
Dr. Gerald L. Esterson, Director, Division of Continuing Professional Education, Washington University
Dr. L. Dale Harris, Associate Dean, College of Engineering, The University of Utah
Dr. James E. Holte, Director, Continuing Education in Engineering and Science, University of Minnesota
Dr. Russell R. O'Neill, Associate Dean and Professor, School of Engineering and Applied Science, University of California at Los Angeles

ENGINEERING PROFESSION ADVISORY GROUP

THEORY AND TECHNIQUES OF OPTIMIZATION FOR PRACTICING ENGINEERS

Raymond L. Zahradnik

Department of Chemical Engineering
Carnegie-Mellon University
Pittsburgh, Pennsylvania

BARNES & NOBLE, INC. NEW YORK

Publishers • Booksellers • Since 1873

L. C. Catalogue Card Number: 70-146261

ISBN 389 00509 6

Distributed

In Canada
by the Ryerson Press, Toronto

In Australia and New Zealand
by Hicks, Smith & Sons Pty. Ltd.,
Sydney and Wellington

*In the United Kingdom, Europe,
and South Africa*
by Chapman & Hall Ltd., London

In Japan
by Maruzen Co. Ltd., Tokyo

Printed in the United States of America

1339449

Preface

There is little need to justify a PECDS text in optimization theory and technique. In our rapidly advancing, highly complex, and extremely competitive technology, it is recognized that success will belong to those who have the ability to make the best decision, quickly and correctly. However, it should be realized that proficiency with various optimizing methods will not guarantee the making of optimum real-world decisions. The effectiveness of a real-world decision, whether it is technical, political, economic or otherwise, depends upon the accuracy of the model used to represent the real-world system and the reasonableness of the criterion used to evaluate the possible alternatives.

To a large extent, one's scientific or engineering training prepares him to frame reasonable descriptions or models of systems to be optimized, but, at best, these models are only approximate. Lack of knowledge of the system, inability to measure or even define key variables, uncertainties in price structures, and inexact identification of system constraints reduce the confidence one places in real-world optimization solutions. However, it is within the framework of engineering uncertainty that engineering decision-making must function. The mathematical theories and techniques of optimization can aid the engineer, scientist, economist or manager in his decision-making, but they cannot serve as substitutes for sound thinking and technical know-how.

Mathematical optimization is a tool, and, like any tool, the more one understands its structure and its functional utility, the more valuable it becomes. Thus, as a user of optimization procedures, one must cultivate an ability to recognize in an assigned engineering — or scientific, or economic, or managerial — situation the existence of an optimization problem, to develop the knowledge to characterize it, and to realize what practical techniques exist to solve it.

In a short text such as this, we cannot hope to completely develop these skills on the part of each reader. We can expect, however, the reader to develop an awareness of the scope of optimization theory, the extent to which practical solution procedures are available, and the limitations on the type and magnitude of the problem that is being solved today. Attention has therefore been limited to the highlights, key-points and workable procedures of optimization, leaving out much of the history, theory and aesthetic appreciation which motivate students spending semesters with the subject. We expect that these aspects of the subject will be self-acquired by the reader as he begins to put optimization to use in his own professional activity.

The concept of optimization, as we know it today, deals with two types of problems. The first kind of problem is one in which values must be assigned to a set of parameters or independent variables to cause some preassigned function of these parameters to be maximized or minimized. Often referred to as "static optimization," this subject will command our attention for the first half of the book. If the function to be maximized or minimized depends not merely on the values of a set of parameters but on their trajectories through some sort of space, we have an entirely different optimization situation. The objective function in this case is known as a functional, because it depends upon the path of the independent variables. These problems are often referred to as dynamic optimization problems and will form the focus for the second half of the book.

The material for this book was assembled for courses given in the chemical engineering department and the post-college professional education program at Carnegie-Mellon University. Several of these were presented in collaboration with Matthew J. Reilly, and his assistance in developing many examples in the text is gratefully acknowledged. The encouragement and support of William W. Ellis, director of the post-college office at Carnegie-Mellon, was instrumental in the completion of this text. Work by LeRoy L. Lynn and Elliot S. Parkin led to many of the examples in Chapter 8. My sincerest appreciation is extended to Dolores Dlugokecki, who typed the manuscript patiently and efficiently.

To Maryann

Contents

PART I
Parameter
Optimization

1
Parameter Optimization

1.1 INTRODUCTION

The theory of ordinary maxima or minima is concerned with the problem of finding the value of an independent variable, x, at which some function of $x - f(x)$ - reaches either a maximum or minimum. Such a point is referred to as an extremum, and the problem of locating the extremum is a parameter optimization problem.

The function f is called the objective function or performance index and may be a representation of many things, depending upon the system chosen — a measure of product quality, life expectancy, sales, operating cost or net profit. Consider it to be a measure by which to judge the effectiveness of the selection of the independent variable(s). It will become apparent that the choice of this measure can be a subtle and sophisticated task, requiring insight into both the system and the optimization procedure.

Initially, much can be learned about the notation, terminology and concepts of the theory of the optimum by examining functions of a single variable. Thus, we first consider single variable optimization.

1.2 SINGLE VARIABLE OPTIMIZATION

A single variable optimization problem is the search for the value of x (call it x^*) which on the closed interval $a \leqslant x \leqslant b$ causes $f(x)$ to be maximized. Here a and b are specific values of x. It is sufficient to seek the maximum of a function, since a minimization problem can always be turned into a maximization one by simply seeking the maximum of $-f(x)$. Moreover, it will always be possible to translate and normalize the x variable so that no loss in generality is incurred if the interval over which $f(x)$ is to be maximized is taken as $[0,1]$. (The closed interval $0 \leqslant x \leqslant 1$ is represented as $[0,1]$. An open interval, represented as $(0,1)$ describes the situation $0 < x < 1$.)

For functions of a single variable, the theorems of calculus provide theoretical relationships and generally useful procedures whereby x^* can be determined. This is not always the case, however, particularly in dealing with functions of many variables. Nonetheless, it is worthwhile to review the major theorems pertaining to the extrema of functions of a single variable in order to establish terminology and to list the exact conditions which prevail at these extrema.

The theorems will be stated without proof; proofs can be found in any text on advanced calculus. The following definitions are useful.

Definition 1.1 By a neighborhood of x^1 is meant an open interval containing x^1. (Superscripts will be used to identify specific values of an independent variable).

Definition 1.2 If f is a function of x defined on the open interval $(0,1)$ and x^1 is a point within the interval, then we say that f has a relative maximum at x^1 if there is some neighborhood of x^1 contained in $(0,1)$, for which $f(x^1) \geqslant f(x)$ where x is a point in the neighborhood.

If the inequality holds strictly, we have a *strong relative maximum*. Otherwise, we have a *weak relative maximum*.

Definition 1.3 f on the interval $[0,1]$ is said to take on its absolute maximum at x^* contained in the interval, if $f(x^*) > f(x)$ for every x on the interval.

In terms of these definitions, we state the following theorem.

Theorem 1.1 Suppose that f has a relative extremum at the point x^* on the open interval $(0,1)$, and suppose that f is differentiable at the point x^*. Then,

$$f'(x*) = 0.$$

Theorem 1.1 states that the vanishing of the first derivative at the extremum is a necessary condition for the extremum.

Example 1.1 Consider the function $f(x) = x^2 - x$ on the interval $[0,1]$. Is the point $x = 1/4$ an extremum? *Solution* $f'(x) = 2x - 1$; $f'(1/4) = 2(1/4) - 1$. Since $f'(1/4) \neq 0$, $x = 1/4$ is not an extremum. The point $x = 1/2$ is an extremum, however, since $f'(1/2) = 2(1/2) - 1 = 0$.

Theorem 1.2 If f is continuous on the interval $[0,1]$ and m and M represent the greatest lower bound and least upper bound of the values of $f(x)$ on this interval, then $f(x)$ assumes each of the values m and M at least once in the interval.

Theorem 1.2 is called the theorem of Weierstrass. It tells us that a continuous f does somewhere attain a maximum. This is an absolute maximum. From Theorems 1.1 and 1.2, it can be concluded that the point at which f attains its absolute maximum is

(a) a point where $f'(x) = 0$, or
(b) a point at one end of the interval, or
(c) a point where f is not differentiable.

Theorem 1.3 Consider the function $f(x)$ defined on the interval $(0,1)$ containing $x*$. Let f be differentiable on the interval, and assume that at $x = x*$ the second derivative exists. If $f'(x*) = 0$, then

(a) if $f''(x*) > 0$, f has a relative minimum at x;
(b) if $f''(x*) < 0$, f has a relative maximum at x;
(c) if $f''(x*) = 0$, no conclusion can be drawn.

In general, if $f'(x*) = f''(x*) = \ldots = f^{(n)}(x*) = 0$, but $f^{(n+1)}(x*) \neq 0$, then if n is odd we have an extremum, depending on the sign. If n is even, we have a horizontal inflection point. This establishes the sufficiency conditions at the extrema.

Example 1.2 The point $x = 1/2$ has been shown to be an extremum for the function $f(x) = x^2 - x$. Does this function have a relative maximum or minimum at $x = 1/2$?

Solution $f'(x) = 2x - 1$, $f''(x) = 2$. Hence, by part (a) of Theorem 1.3, f has a relative minimum at the point $x = 1/2$.

It is advantageous at the onset of an optimization problem to be able to state that a solution exists. We have seen that, if the function to be extremized is continuous and bounded on the closed interval [0,1], then surely the function takes on its maximum value at one or more points on the interval. However, if a point x^1 belonging to the interval [0,1] (which we write as $x^1 \, \epsilon [0,1]$) does satisfy the condition $f'(x^1) = 0$, we must either use Theorem 1.3 or a similar procedure or physical intuition to ascertain whether $f(x^1)$ is a relative minimum, relative maximum or inflection point. Such a procedure is tedious and, particularly in the case of functions of many variables, difficult to apply. However, if we introduce certain types of restrictions on function behavior, several inclusive statements can be made on the nature of the extrema. An extremely useful categorization for this purpose is the notion of convexity and concavity.

Convex Functions A function, $f(x)$, is said to be convex over the interval [0,1] if, for any two points x^1 and x^2 in the interval and for all α, $0 \leqslant \alpha \leqslant 1$,

$$f[\alpha x^2 + (1 - \alpha)x^1] \leqslant \alpha f(x^2) + (1 - \alpha) f(x^1)$$

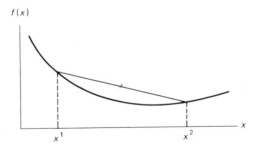

Figure 1.1

— i.e., if the line segment joining any two points of the function lies totally on or above the function itself.

Concave Function   The function $f(x)$ is said to be concave over the interval $[0,1]$ if, for any two points x^1, x^2 in the interval and for all α, $0 \leqslant \alpha \leqslant 1$,

$$f[\alpha x^2 + (1 - \alpha)x^1] \geqslant \alpha f(x^2) + (1 - \alpha) f(x^1)$$

— i.e., the line segment joining any two points of the function lies totally on or below the curve.

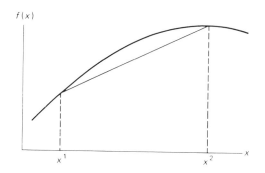

Figure 1.2

It is now possible to state the following theorem:

Theorem 1.4   If f is a convex function over the closed interval $[0,1]$, then any relative minimum of f in the interval is also the absolute (global) minimum of f over the interval.

Example 1.3   Is the function of Example 1 and 2 convex, concave or neither on the interval $[0,1]$?
Solution

$$f[\alpha x^2 + (1 - \alpha)x^1] = [\alpha x^2 + (1 - \alpha)x^1]^2$$

$$- [\alpha x^2 + (1 - \alpha)x^1]$$

$$\alpha f(x^2) = \alpha(x^2)^2 - \alpha x^2$$

$$(1 - \alpha)f(x^1) = (1 - \alpha)(x^1)^2 - (1 - \alpha)x^1$$

It is a simple matter of algebra to show that the difference

$f[\alpha x^2 + (1 - \alpha)x^1] - \alpha f(x^2)$

$\quad - (1 - \alpha) \; f(x^1) = (\alpha^2 - \alpha)(x^1)^2 + (\alpha^2 - \alpha)(x^2)^2$

$\quad + 2\alpha(1 - \alpha)x^1 x^2 = - \alpha(1 - \alpha)(x^1 - x^2)^2.$

Since the last expression is always less than or equal to zero under the restrictions on α, the original difference must also be less than or equal to zero. But this is the condition for a convex function. Hence, we conclude that the relative minimum found at $x = 1/2$ is also the global minimum for $x \epsilon [0,1]$.

Generally, Theorems 1.1-1.4 are easily implemented to yield the solution to a single variable optimization problem. The necessary condition that the first derivative vanish at the extremum is used to identify the point x^*. The use of a necessary condition as a means of locating an extremum is referred to as an indirect method in optimization. This is in contrast to direct methods which locate the extremum by comparing the value of the function at a number of points.

Direct methods find expression in a number of optimum seeking algorithms which prescribe definite procedures for searching the values of a function for its maximum. Such methods will be discussed in a later chapter.

Indirect methods may utilize the necessary conditions for an extremum in two ways. First of all, the condition $f'(x^*) = 0$ may be interpreted as an equation to be solved for x^*. This may be accomplished either analytically or by means of numerical procedures such as the Newton-Raphson technique (2). On the other hand, the condition $f'(x) = 0$ may be used to test whether an arbitrary point, x^1, is an extremum or not by asking if $f'(x^1) = 0$, as was done in Example 1.1. It is this latter philosophy on which most solution techniques are based. If the arbitrary point x^1 does not satisfy the necessary condition for an extremum, a new point, x^2, is selected, based on what is known about the conditions at $x = x^1$, such that x^2 is more likely to satisfy the necessary condition for an extremum than did x^1. The procedures for selecting improved values of the independent variable depends upon the nature of the objective function and, in the case of multivariable problems, upon the nature of the constraint equations as well.

Since solution techniques deal generally with problems involving several independent variables, these ideas are best made quantitative within the context of multivariable optimization.

1.3 MULTIVARIABLE OPTIMIZATION

The theory of ordinary extrema is generalized by considering the problem of determining the values of N independent variables, $x_1, x_2, \ldots x_N$ which maximize a given function of these variables

$$z = f(x_1, x_2, \ldots, x_N). \qquad (1.1)$$

Let the symbol $\bar{x}$ represent the set of independent variables $x_1, x_2, \ldots, x_N$. Then $\bar{x}*$ will represent the particular set of variables, $x_1, x_2, \ldots, x_N$, which maximize z. We can obtain the necessary and sufficient conditions for an extremum by expanding z in a Taylor series about the point $\bar{x}*$. This procedure occurs repeatedly in optimization theory, and it may be well to review what is meant by the expansion. If we have a function of a single variable $f(x)$ such that all of its derivatives exist, we can represent the function at any point x in terms of the function and its derivatives evaluated at the point $x*$ as follows:

$$f(x) = f(x*) + f'(x*)(x - x*) + \frac{f''(x*)}{2!}(x - x*)^2$$
$$+ \ldots + \frac{f^{(k)}(x*)}{k!}(x - x*)^k + \ldots$$

Generally, if we limit our values of x to those close to $x*$, the higher terms of the series may be truncated.

Example 1.4 Expand the function $\sin(x)$ about the point $x = 0$ by using only the first three terms of a Taylor series expansion.

Solution

$$f(0) = \sin(0) = 0$$
$$f'(0) = \cos(0) = 1$$
$$f''(0) = -\sin(0) = 0$$
$$f'''(0) = -\cos(0) = -1$$

Hence,

$$f(x) = \sin(x) = 0 + 1(x - 0) + \frac{0(x - 0)^2}{2!} - \frac{1(x - 0)^3}{3!}$$

$$\sin(x) = x - \frac{x^3}{6}.$$

The Taylor series expansion procedure extends, of course, to a function of many variables, and we can employ it to expand the objective function (1.1) about the point $\bar{x}^*$, keeping only terms of first and second order

$$f(\bar{x}) = f(\bar{x}^*) + \sum_{i=1}^{N} \frac{\partial f}{\partial x_i} (x_i - x_i^*)$$

$$+ \frac{1}{2} \sum_{j=1}^{N} \sum_{i=1}^{N} \left(\frac{\partial^2 f}{\partial x_i \, \partial x_j} \right)(x_i - x_i^*)(x_j - x_j^*). \tag{1.2}$$

Since $f(\bar{x}^*) \geqslant f(\bar{x})$, we can rewrite (1.2) as follows:

$$\sum_{i=1}^{N} \left(\frac{\partial f}{\partial x_i} \right)(x_i - x_i^*)$$

$$+ \frac{1}{2} \sum_{j=1}^{N} \sum_{i=1}^{N} \frac{\partial^2 f}{\partial x_i \, \partial x_j} (x_i - x_i^*)(x_j - x_j^*) \leqslant 0. \tag{1.3}$$

Because $(x_i - x_i^*)$ can have any sign, a necessary condition for (1.3) to be satisfied is that

$$\left(\frac{\partial f}{\partial x_i} \right)_* = 0 \qquad\qquad i = 1, 2, \ldots, N \tag{1.4}$$

A sufficient condition is that the quadratic portion of (1.3) be nonpositive for all values $(x_i - x_i^*)$ and $(x_j - x_j^*)$.

Often the independent variables are related by one or more constraint equations. If there are M such relationships, they may be expressed as follows:

$$g_i(x_1, x_2, \ldots, x_N) = 0 \quad i = 1, 2, \ldots, M < N.$$
$$(1.5)$$

Thus, the extrema are the values of the independent variables $x_1^*, x_2^*, \ldots, x_N^*$ which satisfy (1.5) and which cause z to be maximized. Problems of this sort are illustrated by the following example.

Example 1.5 Find the box of largest volume which can be inscribed in a sphere of unit radius. Let the Cartesian coordinates be labeled as x_1, x_2, x_3.

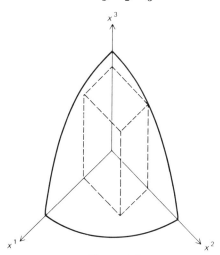

Figure 1.3

The volume of the box $- z -$ is the function to be maximized

$$z = 8 \; x_1 x_2 x_3$$

Subject to the constraint of being inscribed in the unit sphere.

$$x_1^2 + x_2^2 + x_3^2 = 1.$$

Thus, the example includes 3 independent variables, N = 3, a nonlinear objective function, and one nonlinear equality constraint, M = 1.

This problem can be solved in the traditional way of the calculus, and it may be useful to demonstrate this procedure.

Traditional Solution (Differential Approach) Any one of the independent variables may be eliminated from the problem by expressing it in terms of the other two in the constraint equation. For example,

$$x_3 = \sqrt{1 - x_1^2 - x_2^2}$$

$$z = 8x_1 x_2 \left(1 - x_1^2 - x_2^2\right)^{1/2}$$

Equation (1.4) states that a necessary condition x_1 and x_2 must satisfy if they are to maximize z is that the partial derivatives of z with respect to x_1 and x_2 vanish simultaneously at the extremizing points.

$$\left(\frac{\partial z}{\partial x_1}\right)_{x_2} = 8x_2 \left(\sqrt{1 - x_1^2 - x_2^2} - \frac{x_1^2}{\sqrt{1 - x_1^2 - x_2^2}}\right) = 0 \tag{1.6}$$

$$\left(\frac{\partial z}{\partial x_2}\right)_{x_1} = 8x_1 \left(\sqrt{1 - x_1^2 - x_2^2} - \frac{x_2^2}{\sqrt{1 - x_1^2 - x_2^2}}\right) = 0 \tag{1.7}$$

It is easily shown that these equations reduce to the following set:

$$1 - 2x_1^2 - x_2^2 = 0$$

$$1 - x_1^2 - 2x_2^2 = 0$$

from which it follows that $x_1 = x_2 = x_3 = 1/\sqrt{3}$.

The derivatives expressed in (1.6) and (1.7) may be viewed as special derivatives, representing the change of the objective function resulting from perturbations in the $N - M$ variables which remain independent after the M constraint equalities have been incorporated into the objective function. They are called *constrained derivatives* and allow the necessary conditions for unconstrained optimization problems to be applied to problems with constraints. These ideas will be extended to the general problem posed by (1.1) and (1.5) presently, but first an alternate solution procedure for the example is discussed.

Lagrange Multiplier Solution Problems of this type may also be handled by the method of undetermined multipliers, devised in the 18th century by Joseph Louis Lagrange. The method consists of multiplying the constraint equation by an unknown constant, λ, and adding it to the objective function. The new function, F, which is formed in this way is called the Lagrangian and λ is called the Lagrange multiplier.

$$F(x_1, x_2, x_3, \lambda) = 8x_1 x_2 x_3 + \lambda (1 - x_1^2 - x_2^2 - x_3^2)$$

The values of x_1, x_2 and x_3 which maximize the volume of the inscribed box must occur at *stationary points* of the Lagrangian, F. At a stationary point of a function, the partial derivatives of the function with respect to each of the independent variables vanish simultaneously. This means the following equations must hold

$$\frac{\partial F}{\partial x_1} = 8x_2 x_3 - 2 \lambda x_1 = 0$$

$$\frac{\partial F}{\partial x_2} = 8x_1 x_3 - 2 \lambda x_2 = 0 \qquad (1.8)$$

$$\frac{\partial F}{\partial x_3} = 8x_1 x_2 - 2 \lambda x_3 = 0$$

$$\frac{\partial F}{\partial \lambda} = 1 - x_1^2 - x_2^2 - x_3^2 = 0$$

The conditions under which this procedure is valid are relatively mild, as will be demonstrated presently. The advantage to the method is that it makes it unnecessary to solve explicitly for one of the variables in terms of the others (as was done in the first approach) or to take account of the fact that not all variables are independent.

The solution to the set of equations (1.8) is obtained by multiplying the first equation by x_1, the second by x_2 and the third by x_3.

$$8x_1 x_2 x_3 - 2 \lambda x_1^2 = 0$$

$$8x_1 x_2 x_3 - 2 \lambda x_2^2 = 0$$

$$8x_1 x_2 x_3 - 2 \lambda x_3^2 = 0$$

Hence $x_1 = x_2 = x_3$. If we invoke the fourth equation, we obtain the same results as before. Note that the fourth equation is exactly the constraint equation. This is a natural consequence of the method.

We are now in a position to state the general parameter optimization problem and to discuss the conditions which its solution must satisfy. This we will do in the next two sections. The various procedures for solving this problem constitute the subject of the remaining chapters in Part I.

1.4 THE GENERAL PARAMETER OPTIMIZATION PROBLEM

Although the optimization problem defined by (1.1) and (1.5) is reasonably general, there is no transparent provision to account for constraints based on inequalities. For example, in addition to the constraint equations (1.5), it is not uncommon to have additional constraints expressed as inequalities. These may be represented as follows:

$$G_i \ (x_1, \ x_2, \ \ldots \ , \ x_N) \leqslant b_i \ \text{for} \ i \ = \ M \ + \ 1, \ \ldots \ , \ m_1$$
$$(1.9)$$
$$G_i \ (x_1, \ x_2, \ \ldots \ , \ x_N) \geqslant b_i \ \text{for} \ i \ = \ m_1 \ + \ 1, \ \ldots \ , \ m$$
$$(1.10)$$

where the quantities b_i are taken to be non-negative.

In this and subsequent problems, we will assume that the inequality constraints form a convex set. That is, suppose two values of the independent variables $\bar{x}^1$ and $\bar{x}^2$ satisfy constraints (1.9) and (1.10). Then the values of the independent variables formed by a linear combination of $\bar{x}^1$ and $\bar{x}^2$ also satisfy the constraints.

Inequality constraints occur naturally in many engineering and economic contexts, and we will have ample occasion to illustrate them in subsequent chapters. However, for the present, the inequality constraints (1.9) and (1.10) may be converted into equality constraints by defining new variables called slack or surplus variables. Thus the first $m_1 - M$ constraints may be written as

$$x_{si} \ = \ b_i \ - \ G_i \ (x_1, \ x_2, \ \ldots \ , \ x_N) \ \text{or}$$

$$g_i \ (x_1, \ \ldots \ , \ x_N; \ x_{si}) \ = \ G_i \ (x_1, \ \ldots \ , \ x_N) \ + \ x_{si} \ - \ b_i \ = \ 0$$
$$(1.11)$$
$$\text{for} \ i \ = \ M \ + \ 1, \ \ldots \ , \ m_1.$$

The variable x_{si} is termed a slack variable. Clearly the slack variable must satisfy

$$x_{si} \ \geqslant \ 0 \qquad \qquad i \ = \ M \ + \ 1, \ \ldots \ , \ m_1.$$

Likewise the next $m - m_1$ inequality constraints, given by (1.10), are converted into equalities by defining surplus variables.

$$x_{si} \ = \ G_i \ (x_1, \ x_2, \ \ldots \ , \ x_N) \ - \ b_i$$

$$g_i (x_1, \ldots, x_N; x_{si}) = G_i (x_1, \ldots, x_N) - x_{si} - b_i = 0$$

$$i = m_1 + 1, \ldots, m$$

The surplus variable x_{si} must also satisfy non-negativity restrictions

$$x_{si} \geqslant 0 \qquad i = m_1 + 1, \ldots, m$$

It is convenient to make no distinction between the N physically meaningful variables $x_1, x_2, \ldots, x_N$ and the surplus variables x_{xi}. For this purpose, the slack and surplus variables are rewritten as

$$x_{N+i-M} = x_{si} \qquad i = M + 1, \ldots, m.$$

It is further customary to consider that the N original independent variables are required to be non-negative. Although this may appear at first to be somewhat confining, it should be remembered that any value, positive or negative, may be expressed as the difference between two non-negative numbers. Hence the non-negativity restriction on all $n = N + m$ independent variables is customarily employed in the statement of the general parameter optimization problem, which is as follows.

It is desired to determine non-negative values of the n variables $x_1, x_2, \ldots, x_n$ which satisfy the m equations

$$g_i (x_1, x_2, \ldots, x_n) = 0 \qquad i = 1, 2, \ldots, m < n \tag{1.12}$$

and which in addition maximize the function

$$z = f(x_1, x_2, \ldots, x_n). \tag{1.13}$$

Alternatively, it may be required that some or all of the variables be allowed to assume only discrete values, such as integer values. When such constraints are imposed in problems, special solution methods are required which are beyond the scope of this text. However, solution procedures have been worked out in many such instances, and proven algorithms are available.

The general parameter optimization problem is often referred to as a programming problem. This terminology stems from the fact that problems of this sort occur in planning and scheduling contexts and early workers in these fields referred to them as programming problems. Solution techniques are likewise referred to as specific programming techniques — e.g., linear programming.

As mentioned earlier, it is possible to solve the parameter optimization problem posed above by locating the values of the independent variables which satisfy the necessary conditions of the problem. The various solution techniques exploit particular features of the system of equations in determining this location. Before these techniques are discussed, the necessary conditions for the solution of the general parameter optimization problem are derived. These conditions are known as the Kuhn-Tucker conditions after the two mathematicians who developed them in the 1950s (8).

1.5 THE KUHN-TUCKER CONDITIONS

In this section, we will develop in a somewhat heuristic way the Kuhn-Tucker conditions. In this development, we follow closely the derivation presented by Wilde in an early paper (4) and later in his text with Beightler (7).

As in the unconstrained case, we expand the objective function about the point $\bar{x}^*$, keeping, however, only the linear terms

$$f(\bar{x}) - f(\bar{x}^*) = \sum_{j=1}^{m} \left(\frac{\partial f}{\partial x_j} \right) (x_j - x_j^*)$$

$$+ \sum_{j=m+1}^{n} \left(\frac{\partial f}{\partial x_j} \right) (x_j - x_j^*) \tag{1.14}$$

The expansion is written in this way to emphasize that m of the x-variables, in particular the 1st m, are dependent and will be eliminated from consideration by solving the constraint equations. This distinction may be emphasized by introducing different names for the two sets of variables. Let the first m variables be considered as state variables, symbolized as

$$s_j = x_j - x_j^* \qquad j = 1, 2, \ldots, m.$$

The remaining $(n - m)$ variables are considered as decision variables, symbolized as

$$d_j = x_{m+j} - x_{m+j}^* \qquad j = 1, 2, \ldots, (n - m).$$

Equation (1.14) can be written in terms of these variables

$$f(\bar{x}) - f(\bar{x}^*) = \sum_{j=1}^{m} \left(\frac{\partial f}{\partial x_j} \right) s_j + \sum_{j=1}^{n-m} \left(\frac{\partial f}{\partial x_{m+j}} \right) d_j. \tag{1.15}$$

Because the constraint equations are in general nonlinear, their solution could be extremely complex. Therefore, a Taylor series expansion containing only linear terms is used to represent the constraint equations

$$g_i(\bar{x}) - g_i(\bar{x}^*) = \sum_{j=1}^{m} \left(\frac{\partial g_i}{\partial x_j} \right) (x_j - x_j^*)$$

$$+ \sum_{j=m+1}^{n} \left(\frac{\partial g_i}{\partial x_j} \right) (x_j - x_j^*) \qquad i = 1, 2, \ldots, m. \tag{1.16}$$

Since both $\bar{x}$ and $\bar{x}^*$ must satisfy (1.12), we can rewrite (1.16) as follows, in terms of the state and decision variables

$$g_i \, (\bar{x}) \, - \, g_i \, (\bar{x}*) \, = \, \sum_{j=1}^{m} \, \left(\frac{\partial g_i}{\partial x_j} \right) s_j$$

$$+ \, \sum_{j=1}^{n-m} \left(\frac{\partial g_i}{\partial x_{m+j}} \right) d_j \, = \, 0 \qquad i \, = \, 1, \, 2, \, \ldots , \, m.$$

$$(1.17)$$

The m state variables can be expressed in terms of the $(n - m)$ decision variables in a straightforward way; because of the linearity of (1.17), the solution can be expressed in terms of determinants.

Because determinants will be used extensively in the remainder of this development, a brief appendix summarizing their properties is provided at the end of this chapter.

In terms of the quantities appearing in (1.14) and (1.15), we define the following determinants

$$B \, = \, \begin{vmatrix} \dfrac{\partial g_1}{\partial x_1} & \cdots & \dfrac{\partial g_1}{\partial x_m} \\ \vdots & & \\ \dfrac{\partial g_m}{\partial x_1} & \cdots & \dfrac{\partial g_m}{\partial x_m} \end{vmatrix}$$

$$C_i \, = \, \begin{vmatrix} \dfrac{\partial f}{\partial x_{m+i}} & \dfrac{\partial f}{\partial x_1} & \cdots & \dfrac{\partial f}{\partial x_m} \\ \dfrac{\partial g_1}{\partial x_{m+i}} & \dfrac{\partial g_1}{\partial x_1} & \cdots & \dfrac{\partial g_1}{\partial x_m} \\ \vdots & \vdots & & \vdots \\ \dfrac{\partial g_m}{\partial x_{m+i}} & \dfrac{\partial g_m}{\partial x_1} & \cdots & \dfrac{\partial g_m}{\partial x_m} \end{vmatrix} \qquad i \, = \, 1, \, 2, \, \ldots , \, n \, - \, m$$

Determinants such as these, which involve first partial derivatives of functions as the determinant elements are called *Jacobian Determinants.*

It is possible to solve (1.17) for the state variables in terms of the decision variables. Substitution of these solutions into (1.15) yields the following expression for the objection function

$$f(\bar{x}) - f(\bar{x}^*) = \sum_{i=1}^{n-m} \left(\frac{C_i}{B}\right) d_i. \qquad (1.18)$$

Equation (1.18) has the appearance of a Taylor series expansion of f in terms of $d_1, d_2, \ldots, d_{n-m}$ about their zero values. We regard the terms C_i/B as partial derivatives of f with respect to d_i, evaluated at $d_i = 0$. It is useful to consider these as constrained derivatives in the sense that they incorporate the constraints directly into the objective function. If the symbol δ is used to express this idea, the derivatives may be identified as follows:

$$\frac{\delta f}{\delta d_i} = \frac{C_i}{B} \qquad i = 1, 2, \ldots, n - m. \quad (1.19)$$

Equation (1.18) thus becomes

$$f(\bar{x}) - f(\bar{x}^*) = \sum_{i=1}^{n-m} \left(\frac{\delta f}{\delta d_i}\right) di. \qquad (1.20)$$

Since $\bar{x}^*$ was considered to be the point which maximized f, it must be that

$$f(\bar{x}) - f(\bar{x}^*) \leqslant 0.$$

In order for this condition to be satisfied, each of the terms in (1.20) must be nonpositive. For the decision variables, this implies

$$\left(\frac{\delta f}{\delta d_i}\right) d_i \leqslant 0 \qquad i = 1, 2, \ldots, n - m. \tag{1.21}$$

This inequality implies one of two conditions. Since $d_i = x_{m+i} - x^*_{m+i}$, either

(a) $x^*_{m+i} \neq 0$. d_i can be of arbitrary sign. Thus, for (1.21) to be satisfied, it is necessary that $\frac{\delta f}{\delta d_i} = 0$.

(b) $x^*_{m+i} = 0$. Since $x_{m+i} \geqslant 0$, d_i must also be non-negative. Thus $\frac{\delta f}{\delta d_i} \leqslant 0$ is required.

These conditions have a simple geometric interpretation. If the function f incorporating the constraints is plotted as a function of one of the decision variables, d_i,

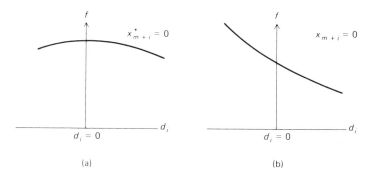

Figure 1.4

holding all the remaining decision variables at their maximizing values, either one of two situations prevails. Since $d_i = x_{m+i} - x^*_{m+i}$, the function achieves a maximum at $d_i = 0$. When $x^*_{m+i} \neq 0$, this maximum is characterized by the vanishing of the constrained derivative (part a of Figure 1.4). When $x^*_{m+i} = 0$, only positive variations of d_i can be considered. Thus, it is only necessary for f to decrease as d_i increases.

Equation (1.21) represents a modified form of what are called the Kuhn-Tucker conditions. They can be used to test a point for relative maximum. If not, then, by using the reasoning employed to derive them, a point can be found where the objective function is improved. This testing procedure is at the bottom of most techniques now available for finding constrained extrema.

The significance of the Kuhn-Tucker conditions is that they provide the necessary conditions for parameter optimization problems involving inequality constraints.

We will now use the procedures developed in this section to resolve the problem posed in example 1.5. Here

$$f = 8x_1 x_2 x_3$$

and

$$g = x_1^2 + x_2^2 + x_3^2 - 1 = 0.$$

Let $s_1 = x_1$ and $d_1 = x_2$, $d_2 = x_3$. Then

$$B = \frac{\partial(g)}{\partial(x_1)} = 2x_1$$

$$C_1 = \begin{vmatrix} \dfrac{\partial f}{\partial x_2} & \dfrac{\partial f}{\partial x_1} \\[2ex] \dfrac{\partial g}{\partial x_2} & \dfrac{\partial g}{\partial x_1} \end{vmatrix} = \begin{vmatrix} 8x_1 x_3 & 8x_2 x_3 \\[1ex] 2x_2 & 2x_1 \end{vmatrix} = \begin{matrix} (8x_1 x_3)(2x_1) \\[1ex] - (8x_2 x_3)(2x_2) \end{matrix}$$

$$C_1 = 16\, x_3\, (x_1^2 - x_2^2)$$

$$C_2 = \begin{vmatrix} \dfrac{\partial f}{\partial x_3} & \dfrac{\partial f}{\partial x_1} \\[2ex] \dfrac{\partial g}{\partial x_3} & \dfrac{\partial g}{\partial x_1} \end{vmatrix} = \begin{vmatrix} 8x_1 x_2 & 8x_2 x_3 \\[1ex] 2x_3 & 2x_1 \end{vmatrix} = \begin{matrix} (8x_1 x_2)(2x_1) \\[1ex] - (8x_2 x_3)(2x_3) \end{matrix}$$

$$C_2 = 16 \ x_2 \ (x_1^2 - x_3^2).$$

The requirement that the constrained derivatives equal zero at the extremum is here expressed as follows, since $x_2^* \neq 0$, and $x_3^* \neq 0$

$$\frac{\delta f}{\delta d_1} = \frac{C_1}{B} = \frac{8x_3}{x_1} \ (x_1^2 - x_2^2) = 0$$

$$\frac{\delta f}{\delta d_2} = \frac{C_2}{B} = \frac{8x_2}{x_1} \ (x_1^2 - x_3^2) = 0.$$

From these equations, it is apparent that $x_1^* = x_2^* = x_3^*$, which fact together with the constraint equation itself allows the problem to be solved.

The Kuhn-Tucker conditions were originally established within the context of the Lagrange multiplier technique. This method, illustrated earlier for example 1.5, consists in forming a Lagrangian function, $F(\bar{x}, \bar{\lambda})$, by multiplying each constraint equation, $g_k = 0$, by an undetermined multiplier, λ_k, $k = 1, 2, \ldots, m$, and adding the resultant sum to f. The symbol $\bar{\lambda}$ depicts the set of the multipliers, $\lambda_1, \lambda_2, \ldots, \lambda_m$.

$$F(\bar{x}, \ \bar{\lambda}) = f(\bar{x}) + \sum_{k=1}^{m} \lambda_k \ g_k \ (\bar{x}) \qquad (1.22)$$

If we neglect for the moment the non-negativity restrictions on the independent variables, the classical necessary condition for an extremum is that it be a stationary point of the Lagrangian; i.e.,

$$\frac{\partial F}{\partial x_1} = \frac{\partial f}{\partial x_1} + \sum_{k=1}^{m} \lambda_k \frac{\partial g_k}{\partial x_1} = 0$$

$$\frac{\partial F}{\partial x_2} = \frac{\partial f}{\partial x_2} + \sum_{k=1}^{m} \lambda_k \frac{\partial g_k}{\partial x_2} = 0 \qquad (1.23)$$

$$\cdot \qquad \cdot \qquad \cdot \qquad \cdot \qquad \cdot$$

$$\frac{\partial F}{\partial x_n} = \frac{\partial f}{\partial x_n} + \sum_{k=1}^{m} \lambda_k \frac{\partial g_k}{\partial x_n} = 0.$$

It is interesting to compare the requirement that the partial derivative of the Lagrangian with respect to the independent variables vanish with the requirement that the constrained derivatives of the objective function vanish. The constrained derivative was identified as

$$\frac{\delta f}{\delta d_i} = \frac{\delta f}{\delta x_{m+i}} = \frac{\begin{vmatrix} \dfrac{\partial f}{\partial x_{m+i}} & \dfrac{\partial f}{\partial x_1} & \cdots & \dfrac{\partial f}{\partial x_m} \\ \dfrac{\partial g_1}{\partial x_{m+i}} & \dfrac{\partial g_1}{\partial x_1} & \cdots & \dfrac{\partial g_1}{\partial x_m} \\ \vdots & & & \\ \dfrac{\partial g_m}{\partial x_{m+i}} & \dfrac{\partial g_m}{\partial x_1} & \cdots & \dfrac{\partial g_m}{\partial x_m} \end{vmatrix}}{\begin{vmatrix} \dfrac{\partial g_1}{\partial x_1} & \cdots & \dfrac{\partial g_1}{\partial x_m} \\ \vdots & & \\ \dfrac{\partial g_m}{\partial x_1} & \cdots & \dfrac{\partial g_m}{\partial x_m} \end{vmatrix}}$$

If we expand the determinant in the numerator in terms of the cofactors of the first column, we observe that for $i = 1, 2, \ldots, (n - m)$

$$\frac{\delta f}{\delta x_{m+i}} = \frac{\partial f}{\partial x_{m+i}} + \sum_{k=1}^{m} \frac{\partial g_k}{\partial x_{m+i}} \frac{A_k}{B} \quad (1.25)$$

where A_k are the cofactors of the first column and are given by

$$A_k = \begin{vmatrix} \dfrac{\partial g_1}{\partial x_1} & \cdots & \dfrac{\partial g_1}{\partial x_m} \\ \vdots & & \\ \dfrac{\partial g_{k-1}}{\partial x_1} & \cdots & \dfrac{\partial g_{k-1}}{\partial x_m} \\ \dfrac{\partial f}{\partial x_1} & \cdots & \dfrac{\partial f}{\partial x_m} \\ \dfrac{\partial g_{k+1}}{\partial x_1} & \cdots & \dfrac{\partial g_{k+1}}{\partial x_m} \\ \vdots & & \\ \dfrac{\partial g_m}{\partial x_1} & \cdots & \dfrac{\partial g_m}{\partial x_m} \end{vmatrix} \quad k = 1, 2, \ldots, m$$

$$(1.26)$$

If we identify the ratio of the determinants A_k/B as λ_k, we may rewrite (1.25) as

$$\frac{\delta f}{\delta x_{m+i}} = \frac{\partial f}{\partial x_{m+i}} + \sum_{k=1}^{m} \lambda_k \frac{\partial g_k}{\partial x_{m+i}} \quad (1.27)$$

$$i = 1, 2, \ldots, n - m.$$

Comparison of (1.27) with (1.24) reveals the constrained derivatives for the variables x_{m+i} to be formally equivalent to the partial derivatives of the Lagrangian with respect to x_{m+i}, $i = 1, 2, \ldots, n - m$.

The Lagrange multiplier technique does not make use of the definition of λ_k; instead, it treats the $\bar{\lambda}$ as unknowns. Moreover it does not eliminate the functional dependency of the Lagrangian on the variables x_1, x_2, $\ldots$, x_m. Hence $2m$ equations are needed in addition to (1.27) to identify the $\bar{x}$ and $\bar{\lambda}$.

The requirement that the derivatives expressed by (1.27) apply to all the independent variables provides m of these conditions. The second set of m equations comes from the constraint equations themselves, which must now be explicitly included since the constraint incorporation provided by the constrained derivative approach has been discarded. The constraints are returned if the derivatives of the Lagrangian function with respect to the Lagrange multipliers are set equal to zero:

$$\frac{\partial F}{\partial \lambda_k} = g_k \ (\bar{x}) = 0 \qquad\qquad k = 1, 2, \ldots, m. \tag{1.28}$$

Hence the necessary conditions for the extremum of the general parameter optimization problem (without the non-negativity restrictions on the $\bar{x}$-vector) are expressed by (1.23) and (1.28) in terms of the Lagrangian.

If non-negativity restrictions are imposed on the $\bar{x}$, the necessary conditions (1.24) and (1.28) can be modified in terms of the constrained derivatives. By utilizing the equivalency between the constrained derivatives and the partial derivatives of the Lagrangian, we may state the following:

For the generalized parameter optimization problem,

Max $z = f(\bar{x})$

subject to $\qquad g_k \ (\bar{x}) = 0 \qquad k = 1, 2, \ldots, m$

and $\qquad\qquad x_i \geqslant 0 \qquad i = 1, 2, \ldots, n > m.$

The solution $\bar{x}^*$ must satisfy the following conditions:

(a) if $x_i^* \neq 0$ $\qquad \dfrac{\partial F}{\partial x_i} = 0$ $\qquad\qquad\qquad\qquad$ (1.29)

$\qquad$ if $x_i^* = 0$ $\qquad \dfrac{\partial F}{\partial x_i} \leqslant 0$ $\qquad i = 1, 2, \ldots, n$

(b) $\qquad\qquad\qquad\qquad \dfrac{\partial F}{\partial \lambda_k} = 0 \quad k = 1, 2, \ldots, m$

$\qquad\qquad\qquad\qquad\qquad\qquad\qquad\qquad\qquad\qquad$ (1.30)

$\qquad$ where $F = f(\bar{x}) + \displaystyle\sum_{k=1}^{m} \lambda_k \, g_k \, (\bar{x})$.

The two parts of condition (a) are sometimes put to-gether in the following way:

$$\left(\frac{\partial F}{\partial x_i} \right) x_i^* = 0 \qquad\qquad (1.31)$$

since either one or other of the terms must be zero. This is known as the complementary slackness principle and must be interpreted as indicated in (1.29).

Just as in the case of functions of a single variable, it is possible to state sufficiency conditions for the general parameter optimization problem. Because these conditions are quite complicated and rarely employed in practice, we will not give them here. The interested reader is referred to references (7) and (11). If certain restrictions are placed on the functions f and g_k, $k = 1, 2, \ldots, m$, how-ever, it is possible to state the sufficient conditions quite simply. The restrictions have to do with the concepts of concavity and convexity introduced in section 1.2 and may be summarized in the following theorem.

Theorem 1.5 Let $f(\bar{x})$ and $g_k(\bar{x})$ $k = 1, 2, \ldots, m$ be concave in $x > 0$. There must be at least one point, $\bar{x}^1$, such that if $\bar{x}^1 \geqslant 0$, $g_{k}(x^1) = 0$, $k = 1, 2, \ldots, m$; i.e., there is at least one feasible solution to the constraint set. Then $\bar{x}^*$ is the optimum solution to the general parameter optimization problem if and only if conditions (1.29) and (1.30) exist. That is, these conditions are both necessary and sufficient.

1.6 APPENDIX: SOME PROPERTIES OF DETERMINANTS

1.6.1 Definition A determinant D is an array of n^2 elements, a_{ij}, arranged in n rows and n columns. Thus,

$$D = \begin{vmatrix} a_{11} & a_{12} & \cdots & a_{1n} \\ a_{21} & a_{22} & \cdots & a_{2n} \\ \vdots & & & \\ a_{n1} & a_{n2} & \cdots & a_{nn} \end{vmatrix}$$

The value of D is obtained by performing certain operations on the elements. This operation consists in summing certain products of the elements, each such product having n terms such that exactly one element is included from each row and column

$$D = \sum (-1)^e \, a_{1k_1} \, a_{2k_2} \cdots a_{nk_n} \quad (1.32)$$

Each product is assigned a multiplier — either $+1$ or -1 — depending upon the order of the second subscripts (provided the product is written as shown, with ordered first subscripts). In general the second subscripts will not be in their natural order $(1, 2, \ldots, n)$. However, they could be restored to their natural order by a finite number of exchanges between two adjacent elements. The number of such exchanges is denoted by e, where e is the exponent on the (-1) multiplier in (1.32). Since e will be either even or odd, the multiplier is either $+ 1$ or $- 1$.

In order to demonstrate how e is obtained, consider the sequence 132. A single exchange of the 2 and 3 returns the sequence to its natural order; hence $e = 1$. The sequence 312 on the other hand requires two exchanges to return it to the order 123, and hence $e = 2$.

The determinant is often referred to as n^{th} order and there are $n!$ terms in its product expansion. This is illustrated for a third-order determinant

$$\begin{vmatrix} a_{11} & a_{12} & a_{13} \\ a_{21} & a_{22} & a_{23} \\ a_{31} & a_{32} & a_{33} \end{vmatrix} \begin{aligned} &= a_{11}a_{22}a_{33} + a_{12}a_{23}a_{31} + a_{13}a_{21}a_{32} \\ &- a_{13}a_{22}a_{31} - a_{12}a_{21}a_{33} - a_{11}a_{23}a_{32} \end{aligned} \qquad (1.33)$$

1.6.2 Minors and Cofactors If m rows and m columns are deleted from an nth order determinant, the remaining elements constitute an $(n - m)^{th}$ order determinant called a minor. Thus, if we were to strike out the first row and column from a third order determinant, the remaining elements would form a second-order determinant,

$$\begin{vmatrix} a_{11} & a_{12} & a_{13} \\ a_{21} & a_{22} & a_{23} \\ a_{31} & a_{32} & a_{33} \end{vmatrix} \qquad \begin{vmatrix} a_{22} & a_{23} \\ a_{32} & a_{33} \end{vmatrix}$$

third order determinant   minor.

When a single row and column are removed from a determinant, the resulting minor with the appropriate sign is known as the cofactor of the element which is at the intersection of the deleted row and column. If the sum of the indices of the element is odd, the sign is +; if the sum of the indices is even, the sign is −. Thus in our example the indicated minor is the cofactor of a_{11} since its sign is correctly indicated as plus.

The utility of cofactors comes from the fact that the value of the determinant may be expressed in terms of the summation of the products of the elements from a particular row or column and their cofactors. For example, the value of a third-order determinant may be expressed in terms of the elements from the first column and their cofactors:

$$\begin{vmatrix} a_{11} & a_{12} & a_{13} \\ a_{21} & a_{22} & a_{23} \\ a_{31} & a_{32} & a_{33} \end{vmatrix} = a_{11} \begin{vmatrix} a_{22} & a_{23} \\ a_{32} & a_{33} \end{vmatrix} - a_{21} \begin{vmatrix} a_{12} & a_{13} \\ a_{32} & a_{33} \end{vmatrix}$$

$$+ a_{31} \begin{vmatrix} a_{12} & a_{13} \\ a_{22} & a_{23} \end{vmatrix}$$

It is easily verified that this equation gives exactly the value obtained in (1.33).

1.6.3 Solution of Linear Simultaneous Algebraic Equations Consider the system of n equations in n unknowns

$$a_{11}x_1 + a_{12}x_2 + \ldots + a_{1n}x_n = b_1$$
$$a_{21}x_1 + a_{22}x_2 + \ldots + a_{2n}x_n = b_2 \quad (1.34)$$
$$\vdots$$
$$a_{n1}x_1 + a_{n2}x_2 + \ldots + a_{nn}x_n = b_n$$

It can be shown that the solution to this system can be expressed in terms of the following determinants

$$B = \begin{vmatrix} a_{11} & a_{12} & \cdots & a_{1n} \\ a_{21} & a_{22} & \cdots & a_{2n} \\ \vdots & & & \\ a_{n1} & a_{n2} & \cdots & a_{nn} \end{vmatrix}$$

$$C_i = \begin{vmatrix} a_{11} & \cdots & a_{1\,i-1} & b_1 & a_{1\,i+1} & \cdots & a_{1n} \\ a_{21} & & a_{2\,i-1} & b_2 & a_{2\,i+1} & & a_{2n} \\ \vdots & & \vdots & \vdots & \vdots & & \vdots \\ a_{n1} & & a_{n\,i-1} & b_n & a_{n\,i+1} & \cdots & a_{nn} \end{vmatrix}$$

$$i = 1, \ldots, n$$

The determinant C_i is obtained by replacing the ith column of B with the column consisting of $(b_1, b_2, \ldots, b_n)$. Then the solution to (1.34) is

$$x_i = \frac{C_i}{B} \qquad i = 1, 2, \ldots, n \qquad (1.35)$$

provided the value of B is not zero. If $B = 0$, the determinant is said to be singular. This implies that two or more of the equations in (1.34) are not linearly independent.

Equation (1.35) is known as Cramer's Rule.

Bibliography

1.2 SINGLE VARIABLE OPTIMIZATION

The theorems and definitions in this section can be found in any introductory calculus text. The reader will find such perspective by consulting the following:

1.   Hancock, H. *Theory of Maxima and Minima.* Dover Publications, Inc. New York, 1960.

Numerical solutions to systems of algebraic equations are discussed in texts on numerical analysis, such as

2.   Lapidus, Leon. *Digital Computation for Chemical Engineers.* McGraw—Hill Book Co. New York, 1962.

1.3 MULTIVARIABLE OPTIMIZATION

In addition to the above references, the theorems pertaining to the nature of an extremum for a function of many variables are discussed in advanced calculus texts such as

3.   Taylor, Angus. *Advanced Calculus.* Ginn and Co., Boston, 1955.

1.4 THE KUHN-TUCKER CONDITIONS

The differential approach is treated in more detail in the following:

4.   Wilde, D.J. Differential calculus in nonlinear programming, *Oper. Res.* 10, pp. 764—773, 1962.

5. Wilde, D.J. A review of optimization theory. *Ind. Eng. Chem.*, 57, pp. 18—31, 1965.

6. Hadley, G. *Nonlinear and Dynamic Programming.* Addison—Wesley Publishing Co., Inc., Reading, Massachusetts, 1964.

7. Wilde, D.J. and Charles S. Beightler. *Foundations of Optimization*, Prentice—Hall, Inc. Englewood Cliffs, New Jersey, 1967.

It is informative to refer to the original paper in this area:

8. Kuhn, H.W. and A.W. Tucker. Nonlinear programming. *Proceedings of the Second Berkeley Symposium on Mathematical Statistics and Probability*, University of California Press, Berkeley, California, pp. 481—492, 1951.

For an alternative and, in a sense, more general statement of the necessary optimality criterion for a mathematical programming problem, see the following:

9. John, F. Extremum problems with inequalities as side conditions. *Studies and Essays, Courant University Volume* (K.O. Friedrichs, O.E. Neugebauer and J.J. Stoker, *eds.*) John Wiley & Sons, Inc., New York, pp. 187—204, 1948.

10. Mangasarian, O.L. and S. Fromovitz. The Firtz John necessary optimality conditions in the presence of equality and inequality constraints. *J. Math. Anal. Appl.*, 17, pp. 37—47, 1967.

Many excellent papers dealing with the theory and interpretation of the Kuhn-Tucker conditions are found in

11. Arrow, K.J. L. Hurwicz and H.U. Zawa. *Studies in
    Linear and Nonlinear Programming*, Stanford Uni-
    versity Press, Stanford, California, 1958.

2
Linear and Quadratic Programming

2.1 INTRODUCTION

A large number of parameter optimization problems can be expressed exclusively in terms of linear functions. For such problems, a powerful solution technique exists. This technique is known as linear programming and has the advantage that the solution to the optimization problem can be obtained in a finite number of steps.

If all of the constraint equations are linear but the objective function contains quadratic terms in the independent variables, the problem is referred to as a quadratic programming problem. In many cases, quadratic programming problems can also be solved in a finite number of steps.

Because solutions to linear programming problems can be obtained efficiently and unequivocally, many engineers seek a linear framework for their problems in preference to other types of models. The result has been that linear programming is one of the most extensively used tools in modern-day optimization practice.

2.2 A SIMPLE EXAMPLE

Many of the fundamental concepts which are used in linear programming are illustrated in the following example.

A small chemicals supplier has received an order for 100 gallons per day of an aqueous solution of a certain chemical. The concentration of this chemical in the desired solution must be at least 5 percent by volume. The supplier does not normally stock a solution of this concentration but may obtain it by mixing together two or more of the standard solutions which he has available. He may also add water to dilute the stronger solutions if that would be desirable. The available solutions and their cost to him are shown below in Table 2.1.

TABLE 2.1

Solution Number, i =	1 (water)	2	3	4	5
Cost, \$/Gal c_i =	0	0.08	0.20	0.36	0.56
Volume Fraction a_i =	0	0.02	0.04	0.06	0.08

The supplier seeks to determine how much of each solution he should use in order to minimize his total cost. The total cost would include labor costs, equipment costs, etc. in addition to the materials costs shown above. Only the latter costs need to be considered in the minimizations, however, since the other costs are fixed and would remain the same regardless of the particular solutions which he used.

The objective function to be minimized may be written

$$z = c_1 x_1 + c_2 x_2 + c_3 x_3 + c_4 x_4 + c_5 x_5$$

$$z = \sum_{i=1}^{5} c_i x_i \tag{2.1}$$

where x_1 is the amount of solution i which is used. This objective function is of a special form and is known as a *linear function*. Linear functions may be identified by the fact that, if all of the variables are multiplied by some number k, then the resulting value of the function is also changed by the multiple of k.

It is necessary that a total of 100 gals of the product solution be prepared. Therefore the variables x_i must satisfy

$$x_1 + x_2 + x_3 + x_4 + x_5 = 100. \qquad (2.2)$$

It is also necessary that the concentration of the product solution be no less than the specified minimum of 5 percent. Thus

$$a_1 x_1 + a_2 x_2 + a_3 x_3 + a_4 x_4 + a_5 x_5 \geqslant 5. \qquad (2.3)$$

Equations (2.2) and (2.3) are the constraints on the problem and express an equality and inequality constraint respectively. Both constraints are linear functions of the variables x_i. Linear programming is capable of handling a large number of constraints, provided that each constraint is linear.

In addition to the above constraints, it is necessary to state some *non-negativity restrictions* in order to ensure that any solution will be physically realistic. Thus

$$x_j \geqslant 0 \ , \ j = 1, \ . \ . \ . \ , 5. \qquad (2.4)$$

Equations (2.1-2.4) are now in the form of a general parameter optimization problem. However, because of the linearity of the equations involved, we will be able to develop a very efficient method (the Simplex Method) to find the minimum (or the maximum) to such a problem. Before doing so, it is worthwhile to consider how one might attack the above problem without any explicit knowledge of linear programming technique.

It should first be noted that constraint (2.3) should be treated as an equality rather than as an inequality. This is obvious from the fact that, if any mixture of solutions satisfies (2.3) as a strict inequality, than one can decrease the cost by reducing proportionately the amount of each solution and adding more water. This is not always the case in linear programming, and one major problem is that of determining when an inequality should be treated as an equality and when it should be treated as a strict in-equality. When an inequality is treated as an equality, it is termed *active*; otherwise, it is termed *inactive*.

It should also be noted that any mixture consisting only of ingredients 1, 2, and 3 would fail to satisfy constraint (2.3) and hence cannot be a solution to the problem. Similarly any mixture consisting only of ingredients 4 and 5 would provide a concentration in excess of 5 percent and hence cannot be an optimal solution to the problem. Thus any feasible solution must consist of at least one ingredient from the first group and at least one from the second group. In general a *feasible solution* to a linear program-ming problem refers to any set of values for the indepen-dent variables x_i which satisfies the constraints and the non-negativity restrictions. It is clear that, in this problem, as is true in most LP problems, there are an infinite number of feasible solutions. Fortunately, only a finite number of these solutions will need to be examined.

Let us return to the problem of searching for an optimal solution by examining the cost of each feasible solution in which only two ingredients are used. For example, let us combine Solutions Number 1 and Number 5. That is, we set

$$x_2 = x_3 = x_4 = 0$$

and choose x_1 and x_5 to satisfy constraints (2.2) and (2.3). They become

$$x_1 + x_5 = 100$$

$$0.08 \ x_5 = 5.$$

Since there are only two equations with two unknowns, a unique solution is possible. One obtains

$$x_1 = 37.5 \qquad x_5 = 62.5.$$

The corresponding cost is calculated from (2.1) and is

$$z = c_1 x_1 + c_5 x_5$$

$$z = 0 + (0.56)(62.5) = \$35.00.$$

By continuing in this fashion and examining all feasible solutions in which all variables but two are zero, we obtain Table 2.2.

TABLE 2.2

Solutions Used	Cost	Amount				
		x_1	x_2	x_3	x_4	x_5
1 and 4	30.00	16.7	--	--	83.3	--
2 and 4	29.00	--	25	--	75	--
3 and 4	28.00	--	--	50	50	--
1 and 5	35.00	37.5	--	--	--	62.5
2 and 5	32.00	--	50	--	--	50
3 and 5	29.00	--	--	75	--	25

The feasible solutions which we have presented in Table 2.2 are of a specified type, known as *basic feasible solutions*. A *basic solution* has no more than m (the number of constraints) variables different from zero. In general, these m *basic variables* are nonzero. The remaining *nonbasic variables* are necessarily zero. In some exceptional cases, however, one or more of the basic variables is also zero, in which case the basic solution is said to be *degenerate*. Degeneracy can cause trouble in any LP problem, and methods for recognizing degeneracy must be exercised.

After examining all of the basic feasible solutions shown in Table 2.2, we could continue the search by considering solutions with three nonzero variables. Actually it is unnecessary to do this since it can be shown that the minimum cost can never be less than the cost for one of the basic feasible solutions. To demonstrate why this would be true, consider a blend of Solutions 2, 3, and 4. That is, set $x_1 = x_5 = 0$. The variables x_2, x_3, and x_4 must satisfy

$$x_2 + x_3 + x_4 \qquad\qquad = 100 \qquad (2.5)$$

$$0.02 \; x_2 + 0.04 \; x_3 + 0.06 \; x_4 = 5. \qquad (2.6)$$

Since there are two equations in three unknowns, we cannot obtain a unique solution. But we can express the nonunique solution in terms of an arbitrary parameter, β. To force such a solution, let us set

$$0.04 \; x_2 = \beta. \qquad (2.7)$$

It can be verified that the solution to (2.5), (2.6), and (2.7) is given by

$$x_2 = 25 \; \beta$$

$$x_3 = 50 \; (1 - \beta)$$

$$x_4 = 25 \; \beta + 50.$$

The reason for expressing the solution in this particular form will become clear shortly. First note, however, that in order to satisfy the nonnegativity restrictions it is necessary that

$$0 \leqslant \beta \leqslant 1.$$

Furthermore, when $\beta = 0$, we have the basic solution which was obtained earlier in terms of the basic variables x_3 and x_4. Similarly, when $\beta = 1$, the basic solution for basic variables x_2 and x_4 is obtained.

The cost for any feasible blend of x_2, x_3, and x_4 may be expressed in terms of the parameter β. Thus

$$z = 0.08 \ x_2 + 0.20 \ x_3 + 0.35 \ x_4$$

$$z = 29\beta + 28(1 - \beta), \ 0 \leqslant \beta \leqslant 1.$$

It is seen that the cost for any blend of these three solutions cannot be less than the cost for a blend of Solutions 3 and 4, nor can it be greater than the cost of a blend of Solutions 2 and 4.

The above result is not surprising since one may arrive at the same conclusion by considering the following process. Two mixtures are prepared, one consisting only of Solutions 2 and 4 and the other containing only Solutions 3 and 4. Each mixture by itself satisfies the concentration constraint, and thus any combination of the two mixtures also satisfies this constraint. The cost of any such combination is a weighed average of the costs of the individual mixtures. In particular,

(Cost of any combination)   = (Cost of mixture from 2 and 4) β

+ (Cost of mixture from 3 and 4) $(1 - \beta)$

where β is relative amount of mixture 2 - 4 which goes into the total. That is

$$\beta = \frac{(\text{amount of mixture from 2 and 4})}{(\text{amount of total combination})} .$$

This example demonstrates the general result that the cost of any nonbasic solution is merely a weighed average of the various basic solutions. It follows that the cost of a non-basic solution cannot be less than the minimum cost of some basic solution.

We now illustrate how this fact can be utilized in order to obtain the solution to an LP problem without having to examine all possible basic feasible solutions. To do this we employ the concept of constrained derivatives and the Kuhn-Tucker Conditions.

2.3 LINEAR PROGRAMMING AND CONSTRAINED DERIVATIVES

Since the objective function and constraint equations are linear, we can express the general parameter optimization problem described in (1.12) and (1.13) as follows

$$\text{Maximize } z = c_1 x_1 + c_2 x_2 + \ldots + c_n x_n$$

$$z = \sum_{j=1}^{n} c_j x_j \tag{2.8}$$

$$\text{subject to } g_i(\overline{x}) = \sum_{j=1}^{n} a_{ij} x_j - b_i = 0 \tag{2.9}$$

$$i = 1, 2, \ldots, m$$

$$\text{and } x_j \geqslant 0 \qquad j = 1, 2, \ldots, n > m \tag{2.10}$$

where the c_j's, a_{ij}'s and b_i's are constants, specified in the statement of the problem.

In order to illustrate the role of the constrained derivatives in the solution of LP problems, consider the following linear programming example.

Example 2.1 Maximize $z = f = 2x_1 + 3x_2$

$$\text{Subject to} \qquad g = 1 - x_1 - x_2 \geqslant 0.$$

$$\text{and} \qquad\qquad x_1 \geqslant 0 \qquad x_2 \geqslant 0.$$

This problem is illustrated in Figure 2.1 which shows the region of feasible solutions and lines of constant f.

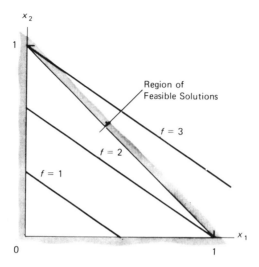

Figure 2.1

Notice that the region of feasible solutions is a closed region in the x_1 x_2 plane. This region is a form of a general *convex polyhedron*, which encloses the region of feasible solutions for a general LP problem. The region has a number of vertices corresponding to intersections of various constraint equations. The vertices represent basic solutions. Hence, by the arguments of the previous section, the optimal solution must occur at one of the vertices of the convex polyhedron enclosing the region of feasible solutions. Let us use this fact to solve the example problem.

Solution This problem is easily solved by inspection of Figure 2.1. However, it is informative to attack the problem with the results of section 1.5. In order to do this the inequality constraint is converted into an equality by introducing a surplus variable x_3 so that the problem becomes

$$z = f = 2x_1 + 3x_2$$

$$1 - x_1 - x_2 - x_3 = 0$$

$$x_1 \geqslant 0, \; x_2 \geqslant 0, \; x_3 \geqslant 0.$$

The vertices represent points where two of the independent variables are zero (in general $n - m$ of the variables will be zero). The nonzero variables (m in general) are the basic variables and form the basic solution or basis.

The linearity of these equations results in all of the unconstrained partial derivatives of f and g being constant, which means that, for each choice of independent variables, the constrained derivatives are constant and independent of the point chosen.

It is clear that the point $x_1 = 0$, $x_2 = 0$, $x_3 = 1$ does not solve the problem. However, consider the point $x_1 = 1$, $x_2 = 0$, $x_3 = 0$. Since $x_2 = 0$ and $x_3 = 0$, by (1.21) the constrained derivatives of f with respect to x_2 and x_3 must be nonpositive. These may be computed as follows:

$$B = \frac{\partial g}{\partial x_1} = -1$$

$$C_2 = \begin{vmatrix} \dfrac{\partial f}{\partial x_2} & \dfrac{\partial f}{\partial x_1} \\[2mm] \dfrac{\partial g}{\partial x_2} & \dfrac{\partial g}{\partial x_1} \end{vmatrix} = \begin{vmatrix} 3 & 2 \\ -1 & -1 \end{vmatrix} = -3 + 2 = 1.$$

Hence $\quad \dfrac{\delta f}{\delta x_2} = \dfrac{C_2}{B} = \dfrac{-1}{-1} = 1$, positive.

Since the constrained derivative of f with respect to x_2 is positive, $x_2 = 0$ is not a solution.

By process of elimination, the third vertex, $x_1 = 0$, $x_2 = 1$, $x_3 = 0$ must be the solution. However, in order to so qualify, $\delta f/\delta x_1$ and $\delta f/\delta x_3$ must be nonpositive. These may be computed also as follows:

$$B = \frac{\partial g}{\partial x_2} = -1$$

$$C_1 = \begin{vmatrix} \dfrac{\partial f}{\partial x_1} & \dfrac{\partial f}{\partial x_2} \\[2mm] \dfrac{\partial g}{\partial x_1} & \dfrac{\partial g}{\partial x_2} \end{vmatrix} = \begin{vmatrix} 2 & 3 \\ -1 & -1 \end{vmatrix} = -2 + 3 = +1$$

$$\frac{\delta f}{\delta x_1} = \frac{C_1}{B} = \frac{+1}{-1} = -1, \text{ nonpositive.}$$

$$C_3 = \begin{vmatrix} \dfrac{\partial f}{\partial x_3} & \dfrac{\partial f}{\partial x_2} \\[2mm] \dfrac{\partial g}{\partial x_3} & \dfrac{\partial g}{\partial x_2} \end{vmatrix} = \begin{vmatrix} 0 & 3 \\ -1 & -1 \end{vmatrix} = 0 + 3 = +3$$

$$\frac{\delta f}{\delta x_3} = \frac{C_3}{B} = \frac{+3}{-1} = -3, \text{ nonpositive.}$$

A generalized procedure for solving the linear programming problem could be developed in this way. However, the linearity permits the use of matrix inversion and multiplication which accomplish the task more efficiently and systematically. Each vertex of the polyhedron enclosing the feasible region represents a point at which $(n - m)$ of the n variables (consisting both of physically meaningful and slack variables) are zero. The linear programming solution algorithm examines these vertices systematically, evaluating the constrained derivatives of each of the variables which is equal to zero. If all of these are nonpositive, the Kuhn-Tucker conditions are satisfied and the problem is solved. If, however, some of the constrained derivatives are positive, then the value of the objective function may be increased by assigning a nonzero value to one of the variables whose constrained derivative was positive.

The value assigned is one which causes another of the independent variables, which originally was nonzero, to become zero. Stated simply, another vertex of the polyhedron is examined.

This procedure ultimately solves the linear programming problem in a finite number of steps and is formalized into the Simplex algorithm which is illustrated in the following section.

2.4 THE SIMPLEX ALGORITHM

The simplex algorithm was developed in 1947 by George Dantzig (5). We illustrate the algorithm first by way of a simple example, following which a more formal statement is made.

Example 2.2 Consider the following linear programming problem.

$$\text{Max} \quad z = 2x_1 + x_2$$

$$\text{Subject to} \quad 4x_1 - x_2 \leqslant 12 \quad \text{(i)}$$

$$x_1 + 2x_2 \leqslant 12 \quad \text{(ii)}$$

$$-3x_1 + 2x_2 \leqslant 4 \quad \text{(iii)}$$

$$x_1 \geqslant 0, \ x_2 \geqslant 0.$$

Figure 2.2 shows graphically the constraints and the region of feasible solutions in the $x_1 - x_2$ plane.

Solution First convert the inequalities to equalities by the addition of slack variables x_3, x_4, and x_5. Thus

$$4x_1 - x_2 + x_3 \qquad\qquad = 12 \quad \text{(i)}$$

$$x_1 + 2x_2 \qquad + x_4 \qquad = 12 \quad \text{(ii)}$$

$$-3x_1 + 2x_2 \qquad\qquad + x_5 = 4. \quad \text{(iii)}$$

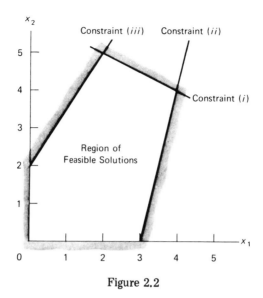

Figure 2.2

Since there are five variables and three constraints, any basis contains three variables, and two nonbasic variables must be set equal to zero. A first choice in this case is to choose the basis which contains x_3, x_4, and x_5 and to set $x_1 = x_2 = 0$.

It is immediately clear that the corresponding basic solution is

$$x_3 = 12$$

$$x_4 = 12$$

$$x_5 = 4$$

and the corresponding value of the objective function is

$$z = 0.$$

There is an alternative method of expressing this same information which has definite advantages. One may merely rewrite the constraints and the objective function and make a note as to which variables are in the basis.

Thus

basis

$$x_3 \quad 12 \; = \; 4x_1 \; = \; x_2 \; + \; x_3 \tag{2.11}$$

$$x_4 \quad 12 \; = \quad x_1 \; + \; 2x_2 \qquad + \; x_4 \tag{2.12}$$

$$x_5 \quad 4 \; = \; 3x_1 \; + \; 2x_2 \qquad\qquad + \; x_5 \tag{2.13}$$

objective function

$$0 \; = \; -2x_1 \; - \; x_2 \qquad\qquad + \; z. \tag{2.14}$$

Note that each constraint contains only one basic variable and that no two constraints contain the same basic variable. Furthermore, since the coefficient of the basic variables contained in any constraint is +1, and since the value of any nonbasic variable is zero, the value of the basic variable is merely the left-hand side of the constraint. Thus the basic feasible solution corresponding to this basis is seen in the left-hand column.

It should also be noted that the objective function is written such that it contains a constant on the left-hand side and only terms involving nonbasic variables on the right-hand side. Since the nonbasic variables are zero, the value of the objective function for this basic solution is the left-hand side.

The coefficients of the objective function, written in this way, are negative values of the constrained derivatives of the objective function, for the nonbasic (zero-valued) variables. If they are all non-negative, the Kuhn-Tucker conditions are satisfied. If any are negative, the conditions are violated and further improvement is possible. That is to say that z can be made larger if either one of the nonbasic variables x_1 or x_2 is made positive. For the sake of definiteness, choose to keep x_2 equal to zero and make x_1 positive since the negative coefficient of x_1 has the greater absolute value.

As x_1 increases, the values of the basic variables x_3, x_4, and x_5 change because the constraints must still be satisfied. In particular, since the coefficients of x_1 in the first two constraints are positive, x_3 and x_4 must decrease as x_1 increases. On the other hand, an increase in x_1 tends to increase x_5. It is clear that, if the coefficients of x_1 in all of the constraints had been negative, then x_1 could have been made arbitrarily large without causing any of the other variables to become negative. In such a case, a finite maximum of the objective function would not exist.

Since in this problem, however, x_3 and x_4 decrease as x_1 increases, it is necessary to determine the maximum value that x_1 can assume without causing x_3 or x_4 to become negative. It is seen from constraint (2.11) that x_3 becomes zero when

$$\frac{12}{4} = 3 = x_1.$$

(Recall that $x_2 = 0$).
From constraint (2.12), one finds that x_4 becomes zero when

$$12 = x_1.$$

Thus, the maximum value that x_1 can assume at this time is 3 since any larger value would force x_3 to become negative.

A new basis is formed which contains x_1 instead of x_3. Furthermore, it is convenient to rewrite the constraints to contain a single different basic variable. To achieve this, divide (2.11) by 4 (the coefficient of x_1), which yields

$$3 = x_1 - 1/4x_2 + 1/4x_3. \tag{2.15}$$

To remove the basic variable x_1 from the second constraint (2.12), multiply (2.11) by $1/4$ and subtract the result from (2.12). The result is

$$9 = 9/4x_2 - 1/4x_3 + x_4. \qquad (2.16)$$

Likewise, x_1 may be removed from the third constraint (2.13) if (2.11) is multiplied by $-3/4$ and subtracted from (2.13).

$$13 = 5/4x_2 + 3/4x_3 + x_5. \qquad (2.17)$$

The basic variable x_1 must also be removed from the right-hand side of the objective function equation. Multiply (2.11) by $-2/4$ and subtract the result from (2.14) to obtain

$$6 = -3/2x_2 + 1/2x_3 + z.$$

It is convenient to collect the several constraints and the objective function in their new forms.

basis

$$x_1 \quad 3 = x_1 - 1/4x_2 + 1/4x_3 \qquad (2.18)$$

$$x_4 \quad 9 = \qquad\quad 9/4x_2 - 1/4x_3 + x_4 \qquad (2.19)$$

$$x_5 \quad 13 = \qquad\quad 5/4x_2 + 3/4x_3 \qquad\quad + x_5 \qquad (2.20)$$

$$6 = \quad z - 3/2x_2 + 1/2x_3 \qquad (2.21)$$

It should be clear that the new basic solution is $x_1 = 3$, $x_4 = 9$, and $x_5 = 13$ and that the objective function has the value of 6.

Note that, since the coefficient of x_2 in the objective function is negative, it is possible to further increase z by making x_2 positive. On the other hand, any positive value of x_3 would cause z to decrease at this step. Thus we form a new basis by adding x_2. One of the old basic variables must be removed, of course, and this will be the one which first becomes zero as x_2 increases. This can only happen if the coefficient of x_2 is positive in the constraint which contains the old basic variable. In this case, either x_4 or x_5 will become zero. From (2.19)

$$x_4 = 0 \text{ when } x_2 = \frac{9}{(9/4)} = 4$$

and from (2.20)

$$x_5 = 0 \text{ when } x_2 = \frac{13}{(5/4)} = \frac{52}{5}.$$

Since x_4 becomes zero first, this variable will be removed from the basis. Consequently, the new basis contains x_1, x_2, and x_5.

The constraint equations must be rewritten such that each one contains a single different basic variable whose coefficient is +1. The following steps must be taken:

(1) divide (2.19) by the coefficient of x_2 in that equation;

(2) multiply (2.19) by a factor which is the ratio of the coefficient of x_2 in (2.18) (that is, $-1/4$) to the coefficient of x_2 in (2.19) (that is, 9/4). Subtract the result from (2.18).

(3) perform a step similar to 2 on (2.20).

(4) perform a step similar to 2 on (2.21).

After carrying out these steps, the following set of equations results:

basis

$$x_1 \quad 4 = x_1 + 2/9x_3 + 1/9x_4$$

$$x_2 \quad 4 = x_2 - 1/9x_3 + 4/9x_4$$

$$x_5 \quad 8 = \qquad 8/9x_3 - 5/9x_4 + x_5$$

$$12 = \qquad 1/3x_3 + 2/3x_4 + z.$$

It is immediately seen that the new basic solution is x_1 = 4, x_2 = 4, and x_5 = 8. The corresponding value of the objective function is $z = 12$. Furthermore, this must be the maximum value of z since the coefficients of the non-basic variables in the objective function equation are both positive. Consequently, if either of these variables takes on a positive value, the objective function must decrease.

It is informative to consider graphically the steps which led to the maximum.

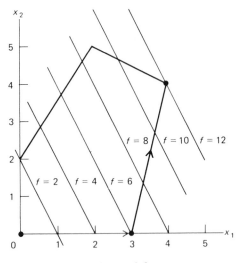

Figure 2.3

Figure 2.3 presents the region of feasible solutions and shows that the first step was from the origin to the neighboring vertex along the x_1 axis. Note from this graph that one could also have moved to the neighboring vertex along the x_2 axis since either move from the origin causes z to increase. At the vertex (3,0), however, one can only move in one direction to the next neighboring vertex. This corresponds to the mathematical observation that at this point only one coefficient in the objective function is negative.

2.5 DUALITY

One of the most interesting properties of linear programming is that there are two distinct ways of formulating the problem. One formulation results in a maximization problem, while the other yields a minimization problem. Each leads to a somewhat different interpretation of the overall situation and is therefore complimentary to the other. By combining these two interpretations, greater understanding and insight into the total problem may be attained. This property of duality is demonstrated in the following example.

Example 2.3 Scott Greenthum, a home gardener, is making plans to fertilize his lawn. After reviewing both the current state of his lawn and the past issue of *American Turf*, a gardening periodical, he has decided that he needs to apply 17 pounds of nitrogen (N), 7 pounds of phosphorus (P), and 5 pounds of potassium (K). Upon making a visit to his local garden supply store, Mr. Greenthum discovers that he cannot obtain his desired blend of ingredients in any single fertilizer. Instead he finds that the four commerical products have the costs and compositions which are shown in Table 2.3.

TABLE 2.3

Available Fertilizers *(Cost per number of pounds)*

	1	2	3	4	Required Amount
Cost	\$5.00/25	\$4.75/25	\$6.25/50	\$4.50/25	
%N	36	22	5	16	17 lbs.
%P	0	8	10	8	7 lbs.
%K	0	8	6	12	5 lbs.

The Primal Problem One way in which Scott might formulate his optimization problem is in terms of minimizing his total cost. If x_i represents the quantity of Fertilizer i which he purchases, then his cost is given by

$$z_p = 0.20x_1 + 0.19x_2 + 0.125x_3 + 0.18x_4 \quad (2.22)$$

where the coefficients express the cost per pound of each fertilizer. To ensure that the combination of fertilizers contains the necessary nutrients, Scott imposes the following inequality constraints.

$$0.36x_1 + 0.22x_2 + 0.05x_3 + 0.16x_4 \geqslant 17 \quad (2.23)$$

$$0.08x_2 + 0.10x_3 + 0.08x_4 \geqslant 7 \quad (2.24)$$

$$0.08x_2 + 0.06x_3 + 0.12x_4 \geqslant 5 \quad (2.25)$$

for the nitrogen, phosphorous, and potassium, respectively. The LP problem is now one of minimizing the objective function (2.22), subject to constraints (2.23), (2.24), and (2.25) and the non-negativity restrictions

$$x_i \geqslant 0.$$

The Dual Problem Alternatively, Scott might express his optimization problem as one of maximizing the total benefit to his lawn. According to this method, he would assign a hypothetical value to each nutrient. Thus, y_1, y_2, and y_3 represent the value per pound of applying the nitrogen, phosphorous, and potassium, respectively. The total value to the lawn would be the sum of the products of the value of each nutrient times the required amount of that nutrient.

$$z_D = 17y_1 + 7y_1 + 7y_2 + 5y_3 \quad (2.26)$$

Note that the total value is not increased if nutrient is used in excess of the required amount.

The nutrient values must be constrained in some fashion since otherwise the total value z_D could be made arbitrarily large. Such constraints may be derived from the statement that the assigned nutrient values must not cause the value of any fertilizer to exceed its cost. The value per pound of any fertilizer is the sum of the nutrient value times the amount of each nutrient which the fertilizer contains. Thus the constraints are

$$0.36y_1 \qquad\qquad\qquad\qquad \leqslant 0.20 \qquad (2.27)$$

$$0.22y_1 + 0.08y_2 + 0.08y_3 \leqslant 0.19 \qquad (2.28)$$

$$0.05y_1 + 0.10y_2 + 0.06y_3 \leqslant 0.125 \qquad (2.29)$$

$$0.16y_1 + 0.08y_2 + 0.12y_3 \leqslant 0.18. \qquad (2.30)$$

The constraints are inequalities rather than equalities since it is possible that the value of some nutrient might be less than its cost. This would clearly be the case if two fertilizers had identical compositions but different costs.

The formulation of the dual problem is completed by specifying that no nutrient value can be negative, i.e.,

$$y_i \geqslant 0 \qquad\qquad i = 1, 2, 3.$$

In view of the fact that both the primal and dual problems arise from the same economic situation, one would expect that a number of relationships would exist between them. This is indeed the case, and one of the most important of these relationships is the fact that the minimum of the primal objective function is equal to the maximum of the dual. That is

$$\text{Min } z_p = \text{Max } z_D. \qquad (2.31)$$

Furthermore, if the Simplex Algorithm is used to solve either of the two problems, then the optimal solution to the other problem can be obtained directly from the final tableau. Duality thus gives us a choice between two alternative methods of solving any linear programming problem. Very often this choice may be made to advantage since one problem may be easier to solve than the other.

The mathematical basis of this duality is found in the Lagrangian statement of the Kuhn-Tucker Theorem. The minimax solution represents the saddle point of the Lagrangian and the y's represent Lagrange multipliers.

We can illustrate this fact by the use of the Lagrange multiplier technique of section 1.5. Consider the generalization of the linear programming problem of example 2.3:

$$\text{Min } z = \sum_{j=1}^{N} c_j x_j \tag{2.32}$$

$$\text{Subject to } \sum_{j=1}^{N} a_{ij} x_j \geqslant b_i \qquad i = 1, 2, \ldots, m \tag{2.33}$$

$$x_j \geqslant 0 \qquad j = 1, 2, \ldots, N \tag{2.34}$$

By the addition of surplus variables, we can transform the inequality constraints to equalities, expressed as follows,

$$\sum_{j=1}^{n} a_{ij} x_j = b_i \qquad i = 1, 2, \ldots, m \tag{2.35}$$

$$x_j \geqslant 0 \qquad j = 1, 2, \ldots, n = N + m. \tag{2.36}$$

Note now that the coefficients a_{ij} for $j = N + 1, \ldots, n$ are given by

$$a_{ij} \begin{cases} = -1 & j = N + i \\ = 0 & \text{elsewhere.} \end{cases}$$

The Lagrangian is now formulated with y_i, $i = 1, 2, \ldots, m$ symbolizing the Lagrange multipliers

$$F(\bar{x}, \bar{y}) = \sum_{j=1}^{n} c_j x_j + \sum_{i=1}^{m} y_i \left(b_i - \sum_{j=1}^{n} a_{ij} x_j \right). \tag{2.37}$$

By the Kuhn-Tucker conditions (1.29)

$$\frac{\partial F}{\partial x_j} = c_j - \sum_{i=1}^{m} y_i a_{ij} \geqslant 0 \qquad j = 1, 2, \ldots, n$$

or

$$\sum_{i=1}^{m} a_{ij} y_i \leqslant c_j. \qquad (2.38)$$

For $j = N + 1, \ldots, n$ (the surplus variables) $c_j = 0$ and a_{ij} given by (2.37) yields the fact that

$$y_i \geqslant 0. \qquad (2.39)$$

Finally, by the complementary slackness principle (1.31), we note that at the extremum

$$\left(\frac{\partial F}{\partial x_j} \right) x_j^* = c_j x_j^* - \sum_{i=1}^{m} y_i a_{ij} x_j^* = 0. \qquad (2.40)$$

Summation of (2.40) yields

$$\sum_{j=1}^{n} c_j x_j^* - \sum_{j=1}^{n} \sum_{i=1}^{m} y_i a_{ij} x_j^* = 0. \qquad (2.41)$$

The first summation is the minimum value of the original objective function. If the order is reversed in the second double summation and equality (2.36) invoked, we obtain

$$\text{Min } z_p = \sum_{j=1}^{n} c_j x_j^* = \sum_{i=1}^{m} b_i y_i^* = \text{Max } z_D. \qquad (2.42)$$

Although we have not proven it here, any values for $\bar{y}$ other than $\bar{y}^*$ will cause $\sum_{i=1}^{m} b_i y_i$ to be less than its value at $\bar{y}^*$.

Hence we seek those values of $\bar{y}$ which

$$\text{maximize} \sum_{i=1}^{m} b_i y_i \qquad (2.43)$$

subject to constraints (2.38) and (2.39). This is the dual to the problem posed by (2.32), (2.33) and (2.34).

If we differentiate the (min z_p) value in (2.42) with respect to b_k, we observe a second role for the dual variables

$$\frac{\partial \ (\text{min} \ z_p)}{\partial \ b_k} = y_k^* \qquad k = 1, 2, \ldots, m \qquad (2.44)$$

Thus y_k^* is the amount the minimum cost would change if a unit change were made in the constraint constant b_k.

Example 2.4 Complete the Example 2.3 and demonstrate the effects of duality.

Solution The final simplex tableau is

$$x_2 \quad 75 = x_2 \qquad \qquad + \quad 2x_1 + 0.67x_4 - 5.6x_5 + \quad 2.8x_6$$

$$x_3 \quad 10 = \qquad x_3 \quad - 1.6x_1 + 0.27x_4 + 4.5x_5 - 12.2x_6$$

$$x_7 \quad 1.6 = \qquad \qquad x_7 - 0.6x_1 - 0.03x_4 + 1.8x_5 - \quad 1.5x_6$$

$$15.50 = z \qquad \qquad - 0.02x_1 - 0.02x_4 - 0.5x_5 - \quad 1.0x_6$$

Hence Scott purchases 75 pounds of fertilizer 2 and 10 pounds of fertilizer 3. His cost is \$15.50. Since $x_7 = 1.6$, he actually gets 1.6 more pounds of potassium than he needs.

The coefficients appearing in the last row are the negatives of the constrained derivatives. They are also the negatives of the dual variables. There is a dual variable associated with every primal constraint and a primal variable associated with every dual constraint. Thus the dual variables associated with x_1, x_2, x_3 and x_4 correspond to the slack variables introduced into the dual constraints, (2.27)-(2.30). Similarly, the primal slack variables correspond to the dual variables appearing in the problem statement.

Thus, if we tabulate the negatives of the coefficients in the objective function row, we find

$$\text{coefficient of} \quad x_1 = 0.02 = y_4$$

$$x_2 = 0.00 = y_5$$

$$x_3 = 0.00 = y_6$$

$$x_4 = 0.02 = y_7$$

$$x_5 = 0.50 = y_1$$

$$x_6 = 1.00 = y_2$$

$$x_7 = 0.00 = y_3$$

The value per pound of nitrogen to Scotts lawn is \$0.50; phosphorous is valued at \$1.00/pound; potassium is valueless. $z_D = \$15.50$.

Application of (2.44) shows that, if the nitrogen requirement could be reduced by 1 pound, Scott's cost would be reduced by \$0.50. A similar reduction of 1 pound in the phosphorous requirement would mean a \$1.00 savings. A one pound reduction in potassium does not alter the cost.

2.6 QUADRATIC PROGRAMMING

The general linear programming problem we have been considering is stated as

$$\text{Max } z = \sum_{j=1}^{n} c_j x_j. \tag{2.45}$$

$$\text{Subject to } \sum_{j=1}^{n} a_{ij} x_j = b_i \quad i = 1, 2, \ldots, m \tag{2.46}$$

$$x_j \geqslant 0 \qquad i = 1, 2, \ldots, n \qquad (2.47)$$

In many cases, the objective function represent a profit, and the $\{c_j\}$ can be looked upon as the selling prices of various commodities. With this point of view, it is reasonable that in some circumstances one can expect the selling prices to reflect quantity discounts. That is, the $\{c_j\}$ factors multiplying the x_j are modified by $q_j x_j$, where q_j is the discount factor per unit. The objective function becomes

$$z = \sum_{j=1}^{n} (c_j - q_j x_j) x_j. \qquad (2.48)$$

Such an index is a quadratic in x_j. Parameter optimization problems which have *quadratic objective functions* and *linear constraints* are known as *quadratic programming problems.*

The quadratic function in (2.48) is a specific form of a more general quadratic function which can be written as

$$\text{Max } z \sum_{j=1}^{n} c_j x_j - \frac{1}{2} \sum_{j=1}^{n} \sum_{i=1}^{n} q_{ij} x_i x_j \qquad (2.49)$$

where without loss of generality we take $q_{ij} = q_{ji}$.

Quadratic objective functions occur naturally in a number of engineering situations. In addition, they may be the result of a statistical analysis of engineering data or a Taylor series expansion of a more complicated function.

In order for the problem to have a finite solution, we require the quadratic term $-\frac{1}{2} \sum_{j=1}^{n} \sum_{i=1}^{n} q_{ij} x_i x_j$ to be nonpositive for all feasible values of x_i and x_j. Under this

restriction, the objective function is concave and the constraints are linear so that the Kuhn-Tucker conditions are
both necessary and sufficient.

Solution procedures for the quadratic programming problem can be developed from either the differential or Lagrange multiplier point of view. Since the two approaches
lead to different algorithms, both will be discussed.

2.6.1 Differential Method Because the constraint
(2.46) equations are linear, it is possible to solve explicitly
for the m state variables in terms of the $n - m$ decision
variables. This solution can be obtained by Cramer's rule.
For example, s_1 is given by

$$s_1 = \frac{\begin{vmatrix} b_1 - \sum\limits_{i=1}^{n-m} a_{1,m+i}\, d_i & a_{12} & \cdots & a_{1m} \\ b_2 - \sum\limits_{i=1}^{n-m} a_{2,m+i}\, d_i & a_{22} & \cdots & a_{2m} \\ \hline \\ b_m - \sum\limits_{i=1}^{n-m} a_{m,m+i}\, d_i & a_{m2} & \cdots & a_{mm} \end{vmatrix}}{\begin{vmatrix} a_{11} & a_{12} & \cdots & a_{1m} \\ a_{21} & a_{22} & \cdots & a_{2m} \\ \\ a_{m1} & a_{m2} & \cdots & a_{mm} \end{vmatrix}} \tag{2.50}$$

By appropriate combination of constants, s_j can be expressed as

$$s_j = \alpha_j + \sum_{i=1}^{n-m} \beta_{ji}\, d_i \tag{2.51}$$

where the $\{\alpha_j\}$ and $\{\beta_{ji}\}$ are obtained by expansion of the determinants appearing in the Cramer's rule solution to (2.46).

Substitution of (2.50) into the objective function results in the following constraint-incorporated form

$$z = \sum_{i=1}^{n-m} \bar{c}_i d_i - \frac{1}{2} \sum_{j=1}^{n-m} \sum_{i=1}^{n-m} \bar{q}_{ij} \, d_i d_j$$

where $\{\bar{c}_i\}$ and $\{\bar{q}_{ij}\}$ are appropriate combinations of the $\{c_i\}$, $\{q_{ij}\}$, $\{\alpha_j\}$ and $\{\beta_{ji}\}$.

$$\frac{\delta z}{\delta d_k} = \bar{c}_k - \sum_{j=i}^{n-m} \bar{q}_{kj} \, d_j \quad k = 1, 2, \ldots, n - m$$
$$(2.52)$$

By the Kuhn-Tucker theorem these derivatives must be non-positive:

$$\text{(a) if } d_k = 0, \frac{\delta z}{\delta d_k} \leqslant 0$$
$$(2.53)$$
$$\text{(b) if } d_k \neq 0, \frac{\delta z}{\delta d_k} = 0$$

A solution procedure may now be outlined.

Step 1. Choose a set of values for the decision variables, $d_1, d_2, \ldots, d_{n-m}$. This selection must be feasible. That is, it must not make any of the state variables negative.

Step 2. Evaluate the constrained derivatives by (2.52). If they satisfy (2.53), the problem is solved. If one or more violate (2.53) choose the decision variable with the most positive constrained derivative. Call this variable d_r.

Step 3.   Change the value of d_r to decrease $\dfrac{\delta z}{\delta d_r}$. Continue this change until one of three events occur:

(a) $\dfrac{\delta z}{\delta d_r}$ becomes zero,

(b) d_r becomes zero, or

(c) one of the state variables, formerly positive, becomes zero.

Step 4.   Depending on the outcome of step 3, we follow three paths:

(a) If $\dfrac{\delta z}{\delta d_r}$ becomes zero, the other constrained derivatives are adjusted by the change $\triangle d_r$ required to make $\dfrac{\delta z}{\delta d_r}$ zero. The Kuhn-Tucker conditions are again tested as in step 2, and the algorithm repeated for a new decision variable if necessary.

(b) If d_r becomes zero, the same procedure used in part (a) above is followed.

(c) If one of the state variables becomes zero, this variable changes places with d_r and the algorithm returns to step 2.

The algebraic manipulations required for these various steps are simple but lengthy. They are straightforward extensions of the operations performed in the solution of linear programming problems by the simplex algorithm. Further details on this method will be found in references (7,8).

2.6.2 Lagrangian Approach

To obtain a solution, we form a Lagrangian function

$$F(\bar{x},\ \bar{\lambda}) = \sum_{j=1}^{n} c_j x_j - \frac{1}{2} \sum_{i=1}^{n} \sum_{j=1}^{n} q_{ij} x_i x_j$$

$$+ \sum_{i=1}^{m} \lambda_i \left(b_i - \sum_{j=1}^{n} a_{ij} x_j \right).$$

Differentiation with respect to x_k gives

$$\frac{\partial F}{\partial x_k} = c_k - \sum_{i=1}^{n} q_{ki}x_i - \sum_{i=1}^{m} \lambda_i a_{ij} \leqslant 0 \tag{2.54}$$

$$k = 1, 2, \ldots, n$$

Equation (2.54) represents n linear inequalities in the $n + m$ variables $x_1, x_2, \ldots, x_n$ and $\lambda_1, \lambda_2, \ldots, \lambda_m$. Solution of these inequalities together with the m equations (2.46) and constraints (2.47) provides the solution to the quadratic programming problem.

The meaning of the inequalities (2.54) must be kept clearly in mind. If $x_k = 0$, then the inequality holds; if $x_k \neq 0$, $\frac{\partial F}{\partial x_k}$ must equal zero.

Inequalities (2.54) may be transformed into equalities by the addition of n slack variables, $w_1, w_2, \ldots, w_n$

$$c_k - \sum_{i=1}^{n} q_{ki}x_i - \sum_{i=1}^{m} \lambda_i a_{ik} + w_k = 0 \tag{2.55}$$

$$k = 1, 2, \ldots, n.$$

In order to satisfy the intention of the Kuhn-Tucker conditions, one or other of the $\{w_k\}$ and $\{x_k\}$ must be zero, i.e.,

$$w_k \, x_k = 0 \qquad\qquad k = 1, 2, \ldots, n \tag{2.56}$$

Equations (2.46) and (2.55) and (2.56) represent a set of $2n + m$ equations in $2n + m$ unknowns. Such a system of equations can be solved by the simplex algorithm of linear programming. In order to do this, we introduce the notion of *artificial* variables.

An artificial variable is a non-negative variable incorporated into the statement of a linear programming problem in order to permit efficient application of the simplex algorithm. In the present context, we have no objective function, only constraint equations. We create an objective function as minus the sum of n non-negative artificial variables, $u_1, u_2, \ldots, u_n$.

$$\text{Max } z = -\sum_{i=1}^{n} u_i. \tag{2.57}$$

The maximum value of z is zero, which occurs when all $\{u_i\}$ are zero. Since the $\{u_i^*\}$ are zero, they can be added to (2.55) in the following way without changing their meaning.

$$c_k - \sum_{i=1}^{n} q_{k\,i} x_i - \sum_{i=1}^{m} \lambda_i a_{ik} + w_k - u_k = 0 \tag{2.58}$$

This completes the solution algorithm.

Summary: the linear programming problem whose solution is also the solution to the quadratic programming problem is

$$\text{Max } z = -\sum_{i=1}^{n} u_i$$

$$\sum_{j=1}^{n} a_{ij} x_j - b_i = 0 \qquad i = 1, 2, \ldots, m$$

$$c_k - \sum_{i=1}^{n} q_{k\,i} x_i - \sum_{i=1}^{m} \lambda_i a_{ik} + w_k - u_k = 0$$

$$w_k x_k = 0, \ w_k \geqslant 0, \ x_k \geqslant 0, \ u_k \geqslant 0$$

$$k = 1, 2, \ldots, n$$

and there may be restrictions on the λ_i if (2.46) were originally inequalities.

Example 2.5

$$\text{Maximize} \quad x_1 + 2x_2 - \frac{1}{2}\,[4x_1^2 - 2x_1 x_2 + x_2^2]$$

$$\text{Subject to} \quad x_1 + x_2 \leqslant 1$$

$$x_1 \geqslant 0, \; x_2 \geqslant 0$$

Solution 1 Change the constraint equation into an equality for solution by the differential algorithm.

$$x_1 + x_2 + x_3 = 1$$

Let $x_3 = s_1$ $\qquad x_1 = d_1$

$$x_2 = d_2$$

so that

$$s_1 = 1 - d_1 - d_2 \tag{i}$$

and

$$\frac{\delta z}{\delta d_1} = 1 - 4x_1 + x_2 \qquad \frac{\delta z}{\delta d_2} = 2 + x_1 - x_2.$$

Consider the feasible point $d_1 = 0$, $d_2 = 0$. This corresponds to the origin ($x_1 = 0$, $x_2 = 0$). At this point, both constrained derivatives are positive.

Hence the Kuhn-Tucker conditions are not satisfied. From step 2, we elect to increase d_2:

(a) we can increase d_2 to 2 before $\dfrac{\delta z}{\delta d_2}$ becomes zero;

(b) we can increase d_2 to 1 before $s_i = 0$.

Hence d_2 can only be increased to 1. From step 4 of the algorithm, s_1 and d_2 must exchange places

$$s_1 = x_2 \qquad d_1 = x_1$$

$$d_2 = x_3$$

$$s_1 = 1 - d_1 - d_2$$

$$z = d_1 + 2(1 - d_1 - d_2) - \frac{1}{2} [4d_1^2 -$$

$$2d_1 (1 - d_1 - d_2) + (1 - d_1 - d_2)^2]$$

$$\frac{\delta z}{\delta d_1} = 1 - 7d_1 - 2d_2$$

$$\frac{\delta z}{\delta d_2} = -1 - 2d_1 - d_2.$$

Our search point is now $d_1 = 0$, $d_2 = 0$. Here $\frac{\delta z}{\delta d_2} \leqslant 0$ so the necessary conditions are satisfied insofar as d_2 is concerned. However, $\frac{\delta z}{\delta d_1} > 0$. Hence we must increase d_1:

(a) we can increase d_1 to the value $1/7$ where $\frac{\delta z}{\delta d_1} = 0$;

(b) at $d_1 = 1/7$, $s_1 = 6/7$.

Hence at $d_1 = 1/7$, $d_2 = 0$, $s_1 = 6/7$, all necessary conditions are satisfied and the problem is solved

$$x_1^* = 1/7, \quad x_2^* = 6/7, \text{ constraint active.}$$

Solution 2 The Lagrange multiplier approach leads to the following linear program.

$$\text{Max } z = -u_1 - u_2$$

subject to

$$x_1 + x_2 + x_3 = 1$$

$$1 - 4x_1 + x_2 - \lambda_1 + w_1 - u_1 = 0$$

$$2 + x_1 - x_1 - \lambda_1 + w_2 - u_2 = 0$$

$$\lambda_1 \geqslant 0, \ x_1 \geqslant 0, \ x_2 \geqslant 0, \ w_1 \geqslant 0, \ w_2 \geqslant 0,$$

$$u_1 \geqslant 0, \ u_2 \geqslant 0$$

$$x_1 w_1 = 0, \ x_2 w_2 = 0.$$

Solution of this problem leads to the final tableau:

$$x_1 \ \ 1/7 = x_1 \qquad\quad + 2/7 \ x_3 - 1/7 \ w_1 + 1/7 \ w_2$$
$$+ \ 1/7 \ u_1 - 1/7 \ u_2$$

$$x_2 \ \ 6/7 = \qquad x_2 \quad + 5/7 \ x_3 + 1/7 \ w_1 - 1/7 \ w_2$$
$$- \ 1/7 \ u_1 + 1/7 \ u_2$$

$$\lambda_1 \ \ 9/7 = \qquad\qquad \lambda_1 - 3/7 \ x_3 - 2/7 \ w_1 - 5/7 \ w_2$$
$$+ \ 2/7 \ u_1 + 2/7 \ u_2$$

$$0 = \qquad\qquad\qquad u_1 + u_2 + z.$$

Generally speaking, the Lagrangian approach leads to a larger problem to solve than the differential approach. However, this problem may be solved by the simplex algorithm (with restricted entry to the basis) and hence should yield the solution in a finite number of steps. The differential algorithm presents a smaller problem to solve. However, it can involve oscillation if two or more constraints are near to each other in the region of feasible solutions. In many engineering problems of practical interest, either approach is satisfactory. For a discussion of several other algorithms for solving the quadratic programming problem, the reader is referred to the reviews in references (4,10).

Bibliography

There are over a thousand possible references to linear programming. A comprehensive treatment of the subject is provided in the following texts, where many additional references can be found.

1. Charnes, A. and W.W. Cooper. *Management Models and Industrial Applications of Linear Programming*, vols. I and II. John Wiley & Sons, New York, 1961.

2. Dantzig, G.B. *Linear Programming and Extensions*. Princeton University Press, Princeton, New Jersey, 1963.

3. Hadley, G. *Linear Programming*. Addison-Wesley Publishing Co., Inc., Reading, Massachusetts, 1962.

2.3 LINEAR PROGRAMMING AND CONSTRAINED DERIVATIVES

The use of constrained derivatives to develop a solution procedure for linear programming problems is discussed in the following:

4. Wilde, Douglass J. and Charles S. Beightler. *Foundations of Optimization*. Prentice-Hall, Inc., Englewood Cliffs, New Jersey, 1967.

2.4 THE SIMPLEX ALGORITHM

The reader will find the early papers of Dantzig of

interest—e.g.:

5. Dantzig, George B. *Programming in a Linear Structure*. Comptroller, USAF, Washington, D. C., 1948.

2.5 DUALITY

The concept of duality is discussed in the references cited above. Further discussion of duality in an operations research context is provided in:

6. Hillier, Frederick S. and Gerald S. Lieberman. *Introduction to Operations Research*. Holden-Day, Inc., San Francisco, California, 1967.

2.6 QUADRATIC PROGRAMMING

Reference (*4*) presents a summary of many of the proposed methods for solving quadratic programming problems. The differential method is generally attributed to Beale.

7. Beale, E.M.L. On minimizing a convex function subject to linear inequalities. *J. Roy. Sta. Soc.*, 17, pp. 172—184, 1955.

8. Beale, E.M.L. On quadratic programming. Nav. Res. Log. Quart., 6, pp. 227—243, 1959.

The Lagrangian Approach employing the simplex method was first suggested by Wolfe.

9. Wolfe, P. The simplex method for quadratic programming. Econ., 27, pp. 382—398, 1959.

The relative merits of these two approaches are discussed by Hadley.

10. Hadley, G. *Nonlinear and Dynamic Programming*. Addison-Wesley Publishing Co., Inc., Reading, Massachusetts, 1964.

3
Geometric Programming

3.1 INTRODUCTION

Geometric programming, like linear and quadratic programming, is an optimization technique which exploits the structure of a parameter optimization problem. The technique is relatively new, being developed in the last decade by Richard Duffin, Clarence Zener and Elmor Peterson (1). The advantages of geometric programming are that it often reduces a complicated optimization problem to one involving a simultaneous set of linear algebraic equations and that it frequently permits one to find the minimum value of an objective function without solving explicitly for the values of the independent variables which yield this minimum. Its disadvantages are that it requires a special form of objective function and constraints.

We can illustrate some of the basic ideas of geometric programming with a simple example.

Example 3.1 A classic example of inventory control is that of the shoeshine boy who specializes in the footwear maintenance of black shoes. Suppose that he is faced with a constant demand for service so that he uses shoe polish at a constant rate of a cans per day. He prefers to purchase his supply of polish in equal numbers of cans, say Q_0 at a time at a cost of c dollars per can. Because he must make the purchase on his own time, he estimates that an hour's worth of sales are lost each time he makes a purchase. This trip to the supplier thus costs him T dollars. At the completion of each day's work, he stores his inventory of polish at the local warehouse at a holding cost of h dollars per day per can.

The problem facing shoeshine boy in this situation (and facing inventory storage generally) is to determine the number of cans of polish he should purchase on each trip to the supplier.

Solution The optimum lot size in this case may be obtained by application of the simple rules of calculus discussed in section 1.2. The cost of raw materials per ordering period, including his trip cost, is given by

$$T + c\,Q_0$$

whereas his storage costs for the period are

$$h \int_0^{t_p} Q(t)\,at.$$

Here t_p denotes the time of a period, and $Q(t)$ denotes the number of cans of polish on hand at any time t. Since demand is constant at a rate a,

$$t_p = Q_0/a.$$

Moreover, at any time t, $0 \leqslant t \leqslant t_p$ the inventory is

$$Q(t) = Q_0 - at.$$

Substitution of the expression for $Q(t)$ into the storage cost expression yields the total storage cost per period: $hQ^2/2a$. Hence, the total cost to shoeshine boy per period z_t is

$$z_t = \frac{T + c\,Q_0 + h\,Q_0^2/2a}{Q_0/a} = ac + \frac{aT}{Q_0} + \frac{h\,Q_0}{2}$$

and the variable portion of the cost z is

$$z = \frac{aT}{Q_0} + \frac{h\,Q_0}{2}.$$

Minimization of this expression yields the well known economic lost size formula

$$Q_0^* = \sqrt{2aT/h}$$

and variable cost $\quad z = \sqrt{2aTh.}$

In order to apply geometric programming to this problem, we must introduce the geometric inequality. If u and v are any two real non-negative variables, the simplest form of the geometric inequality states that the arithmetic mean of the two variables is at least as great as the geometric mean

$$\frac{u}{2} + \frac{v}{2} \geqslant u^{\frac{1}{2}}\,v^{\frac{1}{2}}. \tag{3.1}$$

$$\text{arithmetic mean} \qquad \text{geometric mean}$$

To see this, note that the following inequality is true

$$(u - v)^2 \geqslant 0$$

i.e.,

$$u^2 - 2uv + v^2 \geqslant 0.$$

Add $4uv$ to both sides of the inequality

$$u^2 + 2uv + v^2 \geqslant 4uv$$

or

$$\frac{(u + v)^2}{4} \geqslant uv.$$

Finally

$$\frac{u}{2} + \frac{v}{2} \geqslant u^{\frac{1}{2}} v^{\frac{1}{2}}$$

which is exactly (3.1). Note that the inequality is strict if $u \neq v$. When $u = v$, the inequality becomes an equality.

If we rewrite the variable portion of shoeshine boy's cost in the following way,

$$z = \frac{1}{2} \left(\frac{2aT}{Q_0} \right) + \frac{1}{2} (h \ Q_0).$$

trip cost   holding
cost

Then, by application of the geometric inequality, we find

$$z = \frac{1}{2} \left(\frac{2aT}{Q_0} \right) + \frac{1}{2} (h \ Q_0) \geqslant \left(\frac{2aT}{Q_0} \right)^{\frac{1}{2}} (h \ Q_0)^{\frac{1}{2}}.$$

Moreover we have equality if the two terms are equal. Thus

$$z = \sqrt{2aTh}$$

and

$$2 \left(\frac{aT}{Q_0} \right) = 2 \left(\frac{h \ Q_0}{2} \right)$$

2 x Trip Cost = 2 x Holding Cost

The first expression is identical to the minimum cost obtained earlier; the second equation identifies this minimum as occurring when the trip cost and holding cost are equal.

In this problem, we were fortunate that the independent variable did not appear in the geometric mean of our cost expression. This does not always occur naturally, as can be seen if the holding cost per period has the following form:

$$\text{Holding Cost} = \frac{h \ Q_0^2}{2}.$$

The variable cost term, z, becomes

$$z = \frac{1}{2} \left(\frac{2aT}{Q_0} \right) + \frac{1}{2} \left(h \ Q_0^2 \right) . \qquad (3.2)$$

By (3.1), we find

$$z \geqslant \left(\frac{2aT}{Q_0} \right)^{\frac{1}{2}} \left(h \ Q_0^2 \right)^{\frac{1}{2}} = \sqrt{2aTh \ Q_0};$$

i.e., z depends on Q_0.

Fortunately we can avoid this dilemma by invoking a more general form of the geometric inequality. Thus, in (3.1), consider $u = \dfrac{u_1 + u_2}{2}$ and $v = \dfrac{v_1 + v_2}{2}$.

$$\frac{u_1}{4} + \frac{u_2}{4} + \frac{v_1}{4} + \frac{v_2}{4} \geqslant \left(\frac{u_1 + u_2}{2} \right)^{\frac{1}{2}} \left(\frac{v_1 + v_2}{2} \right)^{\frac{1}{2}}$$

Reapplication of (3.1) to the terms on the right gives the following:

$$\frac{u_1}{4} + \frac{u_2}{4} + \frac{v_1}{4} + \frac{v_2}{4} \geqslant u_1^{\frac{1}{4}} u_2^{\frac{1}{4}} v_1^{\frac{1}{4}} v_2^{\frac{1}{4}}.$$

By similar reasoning, the generalization of (3.1) can be shown to be

$$\delta_1 u_1 + \delta_2 u_2 + \ldots + \delta_\ell u_\ell \geqslant u_1^{\delta_1} u_2^{\delta_2} \ldots u_\ell^{\delta_\ell} \qquad (3.3)$$

where the δ's are arbitrary non-negative weights which satisfy the *normalizing condition*

$$\delta_1 + \delta_2 + \ldots + \delta_\varrho = 1. \tag{3.4}$$

We can apply inequality (3.3) to the variable cost of the modified inventory problem as follows:

$$\delta_1 \left(\frac{aT}{Q_0\delta_1}\right) + \delta_2 \left(\frac{h\ Q_0^2}{2\delta_2}\right) \geqslant \left(\frac{aT}{Q_0\delta_1}\right)^{\delta_1} \left(\frac{h\ Q_0^2}{2\delta_2}\right)^{\delta_2}. \tag{3.5}$$

By (3.4)

$$\delta_1 + \delta_2 = 1.0. \tag{3.6}$$

Since the weights are arbitrary, we can select them to remove the Q_0 dependency from the right-hand side of (3.5). That is,

$$- \delta_1 + 2\delta_2 = 0. \tag{3.7}$$

Equations (3.6) and (3.7) constitute a linear system of equations in two unknowns. Because the number of equations is just equal to the number of unknowns, a unique solution for the δ's is possible. When this is the case, the problem is said to be of zero degree of difficulty. We will discuss in the next section the general conditions for a problem to have zero degree of difficulty. Here the solution of (3.6) and (3.7) is

$$\delta_1 = 2/3, \ \delta_2 = 1/3 \tag{3.8}$$

$$z = \frac{3}{2} \left(a^2 T^2 h\right)^{1/3} \tag{3.9}$$

$$\left(\frac{aT}{Q_0}\right) = 2 \left(\frac{hQ_0^2}{2}\right). \tag{3.10}$$

Here the optimum policy is to make the trip cost equal to twice the holding cost. If this is accomplished, in an optimum way, the variable cost is given by (3.9). Notice that we have obtained this information by solving a linear set of equations and without obtaining an explicit expression for Q_0^*.

3.2 GENERAL STATEMENT OF GEOMETRIC PROGRAMMING TECHNIQUE

In order to apply the ideas of the previous section to problems of a general nature, we must restrict our attention to finding the minimum of objective functions consisting of posynomials.

Definition 3.1 A posynomial, z, is a function of n independent variables $\bar{x} = x_1, x_2, \ldots, x_n$ expressed as the sum of m individual terms consisting of power functions of the independent variables. That is,

$$z = c_1 p_1(\bar{x}) + c_2 p_2(\bar{x}) + \ldots + c_m p_m(\bar{x}) = \sum_{i=1}^{m} c_i p_i(\bar{x}) \tag{3.11}$$

where c_i is a positive constant and $p_i(\bar{x})$ is a power function given by

$$p_i(\bar{x}) = x_1^{a_{i1}} x_2^{a_{i2}} \ldots x_n^{a_{in}} = \prod_{j=1}^{n} x_j^{a_{ij}}. \tag{3.12}$$

It is convenient to regard the objective function expressed in this way as the total cost of a process in which the $\bar{x}$ are independent variables. With this interpretation, the individual terms, $c_i p_i(\bar{x})$ represent component costs, and the independent variables are taken to be nonnegative:

$$x_i \geq 0 \qquad i = 1, 2, \ldots, n. \tag{3.13}$$

We can apply the general form of the geometric inequality, (3.3), to the minimization of the objective function given by (3.11) by identifying $u_i = c_i p_i(\bar{x})/\delta_i$. Then

$$z \geqslant \left(\frac{c_1 p_1}{\delta_1}\right)^{\delta_1} \left(\frac{c_2 p_2}{\delta_2}\right)^{\delta_2} \dots \left(\frac{c_m p_m}{\delta_m}\right)^{\delta_m}. \quad (3.14)$$

The right hand side of (3.14) is termed the pre-dual
function, $V(\bar{x}, \bar{\delta})$ where $\bar{\delta}$ symbolizes the variables δ_1, δ_2,
..., δ_m. Just as in the inventory examples, we will select
$\bar{\delta}$ to eliminate the dependency of the pre-dual function on
the x-variables. In order to do this, we will utilize the
form of the power functions which appear in the expression for the pre-dual function:

$$V(\bar{x}, \bar{\delta}) = \prod_{i=1}^{m} \left(\frac{c_i}{\delta_i}\right)^{\delta_i} \prod_{i=1}^{m} [p_i(\bar{x})]^{\delta_i}. \quad (3.15)$$

But

$$\prod_{i=1}^{m} [p_i(\bar{x})]^{\delta_i} = \prod_{i=1}^{m} \prod_{j=1}^{n} \left(x_j^{a_{ij}}\right)^{\delta_i} = \prod_{j=1}^{n} x_j^{\left[\sum_{i=1}^{m} a_{ij} \delta_i\right]}. \quad (3.16)$$

If each of the exponents on the x-variables is forced to
be zero, the product of the power functions given by (3.16)
is one, and the pre-dual function becomes independent of
$\bar{x}$. In order to achieve this, we require

$$\sum_{i=1}^{m} a_{ij}\delta_i = 0 \qquad j = 1, 2, \dots, n. \quad (3.17)$$

These conditions are referred to as the *orthogonality
conditions*. If they are satisfied, inequality (3.14) becomes

$$z \geqslant \left(\frac{c_1}{\delta_1}\right)^{\delta_1} \left(\frac{c_2}{\delta_2}\right)^{\delta_2} \cdots \left(\frac{c_m}{\delta_m}\right)^{\delta_m} = \prod_{i=1}^{m} \left(\frac{c_i}{\delta_i}\right)^{\delta_i}.$$
(3.18)

The function on the right is termed the dual function and denoted by $v(\bar{\delta})$.

It can be shown that the maximum of $v(\bar{\delta})$ subject to (3.4) and (3.17) is equal to the minimum of z with respect to $\bar{x}$.

When the δ's are uniquely determined by (3.4) and (3.17), we have a problem with zero degree of difficulty. This occurs when the number of equations, $n + 1$, equals the number of terms in the objective function, m. If m is greater than $n + 1$, the remaining conditions on $\bar{\delta}$ come from the minimizing requirement on $v(\bar{\delta})$, and the problem is said to have degree of difficulty $m - n - 1$. If m is less than $n + 1$, the problem has no finite solution.

The values of $\bar{\delta}$ which are computed, either uniquely from the constraints for a problem of zero degree of difficulty or from maximizing the dual function subject to the constraints, are symbolized by $\bar{\delta}*$. The value of the dual function evaluated at $\bar{\delta}*$ is the minimum of the primal function

$$\min z = v(\bar{\delta}*).$$

Moreover, the individual components of z, as we have seen, contribute a fraction to the minimum cost equal to their respective weights, i.e.,

$$c_i p_i(\bar{x}*) = \delta_i^* \, v(\bar{\delta}*) \qquad i = 1, 2, \ldots, m.$$
(3.19)

Example 3.2 Multistage Compression Consider the case of the adiabatic, reversible 3-stage compression of a mole of ideal gas, with constant ratio of specific heat at constant pressure to specific heat at constant volume. If there is perfect intercooling between stages, the work required for compression ($- W$) is given by (4)

$$- W = \frac{p_1 V_1}{\epsilon} \left[\left(\frac{p_2}{p_1}\right)^{\epsilon} + \left(\frac{p_3}{p_2}\right)^{\epsilon} + \left(\frac{p_4}{p_3}\right)^{\epsilon} - 3 \right]$$

where

p_1 = inlet pressure; p_4 = outlet pressure

V_1 = inlet molar volume

p_2, p_3 = intermediate pressures to be selected

ϵ = $(k - 1)/k$ where k is ratio of specific heats.

We can define a new index, z, as follows

$$z = -\frac{W\epsilon}{p_1 V_1} + 3 = \left(\frac{1}{p_1}\right)^{\epsilon} p_2^{\epsilon} + \frac{p_3^{\epsilon}}{p_2^{\epsilon}} + \left(p_4^{\epsilon}\right) \frac{1}{p_3^{\epsilon}}.$$

Clearly, minimizing z also minimizes the work required for compression. Note that z is a posynomial and hence can be minimized by the geometric programming technique. The dual function is

$$v(\bar{\delta}) = \left[\left(\frac{1}{p_1}\right)^{\epsilon} \frac{1}{\delta_1}\right]^{\delta_1} \left(\frac{1}{\delta_2}\right)^{\delta_2} \left(\frac{p_4^{\epsilon}}{\delta_3}\right)^{\delta_3}.$$

The orthogonality conditions are

$$p_2: \quad \epsilon\, \delta_1 - \epsilon\, \delta_2 = 0$$

$$p_3: \quad \epsilon\, \delta_2 - \epsilon\, \delta_3 = 0.$$

The normality condition is

$$\delta_1 + \delta_2 + \delta_3 = 1.$$

The problem has zero degree of difficulty, and it is obvious that the value of the weights is

$$\delta_1^* = \delta_2^* = \delta_3^* = 1/3.$$

This indicates that, in a three-stage compressor, each stage should do equal work for optimum operation. This result is, of course, well known (*4*) and has recently been derived within the geometric programming context by R. I. Kermode (*3*).

Example 3.3 Optimum Design Duffin, Zener and Peterson have presented many examples in their text on geometric programming; the following is a variation on one of these.

Suppose that an open cylindrical vessel is to be constructed to transport 10 cubic feet of liquid from one location in a processing plant to another. The sides of the vessel can be made from material costing $1.00 per square foot, but the bottom must be made from a special alloy costing $10.00 per square foot. If the cost per round trip is 10 cents and the container will have no salvage upon completion of the operation, what is the minimum cost of moving the liquid?

Solution The objective function in this case is the cost of the operation in dollars:

$$z = \frac{(10)(.1)}{\pi r^2 h} + 1(2\pi rh) + 10(\pi r^2)$$

$$\underbrace{\qquad\qquad}_{\text{Operating}\atop\text{Cost}} \qquad \underbrace{\qquad\qquad\qquad}_{\text{Fixed or Capital Investment}\atop\text{Cost}}$$

where r = radius of vessel (ft)

h = vessel height   (ft).

The dual function is

$$v(\bar{\delta}) = \left(\frac{1}{\pi\delta_1}\right)^{\delta_1} \left(\frac{2\pi}{\delta_2}\right)^{\delta_2} \left(\frac{10\pi}{\delta_3}\right)^{\delta_3}$$

We seek weights which satisfy the orthogonality conditions

$$r: \quad -2\delta_1 + \delta_2 + 2\delta_3 = 0$$

$$h: \quad -\delta_1 + \delta_2 \quad\quad = 0$$

and the normality condition

$$\delta_1 + \delta_2 + \delta_3 = 1.$$

The problem has again zero degree of difficulty; the weight values are

$$\delta_1^* = 2/5 \;, \quad \delta_2^* = 2/5 \;, \quad \delta_3^* = 1/5.$$

This says that the operating costs should constitute 2/5 of the project cost. The cost of the sides of the vessel should be 2/5 and the bottom 1/5. The cost of the operation is obtained by substitution of the weight values into the dual function

$$v(\bar{\delta}^*) = \left(\frac{5}{2\pi}\right)^{2/5} \left(\frac{10\pi}{2}\right)^{2/5} \left(50\pi\right)^{1/5}$$

$$v(\bar{\delta}^*) = \$7.55.$$

In order to obtain the values of the independent variables in the primal problem, namely r^* and h^*, we once again exploit the nature of the posynomial objective function. After rearranging (3.19), we note that

$$p_i(\bar{x}^*) = \frac{v(\bar{\delta}^*)\; \delta_i^*}{c_i} \quad\quad i = 1, 2, \ldots, m.$$

Each term on the right hand side of this equation is known, so that $p_i(\bar{x}^*)$ can be found. Substitution of (3.12) into (3.19) yields

$$\prod_{j=1}^{n} x_j^{*^{a_{ij}}} = \left[\frac{v(\bar{\delta}^*)\; \delta_i^*}{c_i}\right] \quad\quad i = 1, 2, \ldots, m. \tag{3.20}$$

Taking the logarithm of (3.20) yields

$$\sum_{j=1}^{n} a_{ij} \, \ell\eta \, x_j^* = \ell\eta \left[\frac{v(\bar{\delta}^*) \, \delta_i^*}{c_i} \right] \quad i = 1, 2, \ldots, m. \tag{3.21}$$

For convenience, define

$$w_j^* = \ell\eta \, x_j^*.$$

Then (3.21) can be written as

$$\sum_{j=1}^{n} a_{ij} w_j^* = \ell\eta \left[\frac{v(\bar{\delta}^*) \, \delta_i^*}{c_i} \right] \quad i = 1, 2, \ldots, m. \tag{3.22}$$

Equation (3.22) is a set of m linear equations in n unknowns (w_j^* $j = 1, 2, \ldots, n$). Since $m > n$, we have more equations than unknowns. The theorems of geometric programming assure us that a unique solution can be found. The techniques for solving systems of linear equations from chapter 1 can be used.

Returning now to example 3.3, we employ (3.21). Let

$$w_1 = \ell\eta \, r$$

$$w_2 = \ell\eta \, h$$

$$-2 \, w_1^* - w_2^* = \ell\eta \left[\frac{(7.55)(2/5)}{(1/\pi)} \right] = \ell\eta \, [9.6] = 2.25$$

$$w_1^* + w_2^* = \ell\eta \left[\frac{(7.55)}{2\pi} \left(\frac{2}{5} \right) \right] = \ell\eta \, [0.48] = -0.74$$

$$2 \, w_1^* = \ell\eta \left[\frac{(7.55)}{10\pi} \left(\frac{1}{5} \right) \right] = \ell\eta \, [0.048] = -3.02.$$

The solution to these equations, which may be obtained from any two, is

$$w_1^* = -1.51 \qquad w_2^* = 0.77.$$

Hence

$$r^* = e^{-1.51} = 0.22 \text{ ft.}$$

$$h^* = e^{\,0.77} = 2.16 \text{ ft.}$$

Example 3.4 Suppose now the vessel height in example 3.3 is set at 1.0 feet. The objective function is

$$z = \frac{1}{\pi r^2} + 2\pi r + 10\pi r^2.$$

Since the height is not set at its optimum value, the minimum value of z in this case must be greater than \$7.55; i.e.,

$$7.55 < \text{Min } z.$$

Although this fact is obvious, it is interesting to note that in this case the dual function is identical to the one from example 3.3. Hence if we pick any nonoptimum set of values for the weights and use them to evaluate the dual function, the value obtained is less than the minimum of z.

Similarly any nonoptimum r value would give a value to the objective function greater than its minimum. For example, if we use the previous result, $r = 0.22$, the present objective function is 9.46. Hence,

$$7.55 < \text{Min } z < 9.46.$$

In many cases, the ability to bracket the answer may be all that is desired, and the duality theory of geometric programming can be used to provide this bracket.

We can, however, proceed formally with the geometric

programming algorithm to obtain a solution. Since $n = 1$ and $m = 3$, we have a problem of $m - (n + 1)$ or 1 degree of difficulty. The orthogonality condition is

$$- 2\ \delta_1 + \delta_2 + 2\ \delta_3 = 0$$

and the normality condition is

$$\delta_1 + \delta_2 + \delta_3 = 1.$$

From these equations, we can express δ_2 and δ_3 in terms of δ_1

$$\delta_2 = 2 - 4\delta_1$$

$$\delta_3 = 3\delta_1 - 1.$$

If these expressions are substituted into the dual function, the following results

$$v(\bar{\delta}) = \left(\frac{1}{\pi\delta_1}\right)^{\delta_1}\left(\frac{2\pi}{2 - 4\delta_1}\right)^{(2 - 4\delta_1)}\left(\frac{10\pi}{3\delta_1 - 1}\right)^{(3\delta_1 - 1)}$$

It is obvious in this case that maximizing the dual function is a problem of equal or greater complexity than minimizing the original objective function. No advantage has been gained by application of the geometric programming algorithm. This will generally be the case when the degree of difficulty is equal to the number of original independent variables.

One can show that $v(\delta)$ above is maximized for $\delta_1 = 0.44$; $\delta_2 = 0.225$ and $\delta_3 = 0.335$. $v(\delta^*) = 8.22$ and $r^* = 0.295$.

While the examples have pointed out the strengths and weaknesses of geometric programming as applied to the minimization of a posynomial, nothing has been said about the constrained optimization of such functions. Fortunately, if the constraints are posynomial inequalities, the method can be used in much the same form as demonstrated in this section.

3.3 HANDLING CONSTRAINTS

In order to illustrate the type of constraints which can be handled by geometric programming, consider the following example.

Example 3.5 Suppose that the design problem discussed in example 3.3 had been phrased in the following way. Minimize the cost of *constructing* the cylindrical vessel subject to the constraint that only 10 trips are allowed. That is,

$$\text{Min } z = 2\pi rh + 10\pi r^2$$

$$\text{S.T. } \frac{10}{\pi r^2 h} \leqslant 10$$

or

$$\frac{1}{\pi r^2 h} \leqslant 1.$$

Geometric programming can handle the problem posed in this example. In general, it can handle the problem of minimizing a posynomial subject to r constraints, if the constraints are expressible in the following way:

$$g_k(\bar{x}) \leqslant 1 \qquad k = 1, 2, \ldots, r$$

where the functions $g_k (\bar{x})$ $k = 1, 2, \ldots, r$ are posynomials. Let us label the components in these posynomials in a sequential fashion. Since there were m components in the objective function, we would let

$$g_1(\bar{x}) = c_{m+1} p_{m+1}(\bar{x}) + \ldots + c_{m_1} p_{m_1}(\bar{x}) \tag{3.23}$$

$$g_2(\bar{x}) = c_{m_1+1} \, p_{m_1+1}(\bar{x}) + \ldots + c_{m_2} \, p_{m_2}(\bar{x})$$

$$-\quad-\quad-\quad-\quad-\quad-\quad-$$

$$g_r(\bar{x}) = c_{m_{r-1}+1} \, p_{m_{r-1}+1}(\bar{x}) + \ldots + c_{m_r} \, p_{m_r}(\bar{x}).$$

Consider the case where $r = 1$ — i.e., a single constraint. Since $g_1(\bar{x})$ is a posynomial, the following version of the geometric inequality applies

$$g_1(\bar{x}) \geqslant \left(\frac{c_{m+1} \, p_{m+1}(\bar{x})}{\delta_{m+1}}\right)^{\frac{\delta_{m+1}}{\lambda_1}} \ldots \left(\frac{c_{m_1} \, p_{m_1}(\bar{x})}{\delta_{m_1}}\right)^{\frac{\delta_{m_1}}{\lambda_1}} \lambda_1 \tag{3.24}$$

where $\quad \delta_{m+1} + \delta_{m+2} + \ldots + \delta_{m_1} = \lambda_1.$ $\quad$ (3.25)

In the case of constraints, the weights are not normalized but instead set equal to a constant λ_1. This introduces an extra term λ_1 on the right hand side of the inequality (3.24).

In terms of the parameter optimization theory of chapter 1, λ_1 may be looked upon as a Lagrange multiplier, used to append the constraint equation to the objective function. It can be shown that λ_1 must be non-negative. Moreover, if λ_1 is nonzero at the extremum, the constraint is active (i.e., an equality); if λ_1 is zero, the constraint is inactive (i.e., $g_1(\bar{x}^*) < 1$).

To see how (3.24) comes about, consider the geometric inequality applied to $g_1(\bar{x})$ with normalized weights $\triangle_{m+1}$, $\triangle_{m+2}, \ldots, \triangle_{m_1}$

$$\left[g_1(\bar{x})\right] \geqslant \left(\frac{c_{m+1} \, p_{m+1}(\bar{x})}{\triangle_{m+1}}\right)^{\triangle_{m+1}} \ldots \left(\frac{c_{m_1} \, p_{m_1}(\bar{x})}{\triangle_{m_1}}\right)^{\triangle_{m_1}}$$

$$\tag{3.26}$$

where $\triangle_{m+1} + \triangle_{m+2} + \ldots + \triangle_{m_1} = 1.$ (3.27)

Comparison of (3.27) with (3.25) shows

$$\triangle_k = \frac{\delta_k}{\lambda_1} \qquad\qquad k = m + 1, \ldots, m_1. \quad (3.28)$$

Substitution of this value for the $\triangle_k$'s in (3.26) yields inequality (3.24).

Since $g_1(\bar{x}) \leqslant 1$, inequality (3.24) can be written as

$$1 \geqslant \left(\frac{c_{m+1}\, p_{m+1}(\bar{x})}{\delta_{m+1}} \right)^{\dfrac{\delta_{m+1}}{\lambda_1}} \ldots \left(\frac{c_{m_1}\, p_{m_1}(\bar{x})}{\delta_{m_1}} \right)^{\dfrac{\delta_{m_1}}{\lambda_1}} \lambda_1. \quad (3.29)$$

If inequality (3.29) is raised to the λ_1 power,

$$1 \geqslant \left(\frac{c_{m+1}\, p_{m+1}(\bar{x})}{\delta_{m+1}} \right)^{\delta_{m+1}} \ldots \left(\frac{c_{m_1}\, p_{m_1}(\bar{x})}{\delta_{m_1}} \right)^{\delta_{m_1}} {}^{\lambda_1}_{\lambda_1}. \quad (3.30)$$

Inequality (3.30) is multiplied by inequality (3.14), which involves the objective function, to yield

$$z \geqslant \left(\frac{c_1\, p_1}{\delta_1} \right)^{\delta_1} \ldots \left(\frac{c_m\, p_m}{\delta_m} \right)^{\delta_m} \left(\frac{c_{m+1}\, p_{m+1}}{\delta_{m+1}} \right)^{\delta_{m+1}} \quad (3.31)$$

$$\ldots \left(\frac{c_{m_1}\, p_{m_1}}{\delta_{m_1}} \right)^{\delta_{m_1}} {}^{\lambda_1}_{\lambda_1}.$$

where $\delta_1 + \delta_2 + \ldots + \delta_m = 1$ (3.32)

$$\delta_{m+1} + \delta_{m+2} + \ldots + \delta_{m_1} = \lambda_1. \quad (3.33)$$

If we have more than one constraint, the additional terms are introduced into inequality (3.31) in a similar fashion, a constant λ_k being included for each of the constraints.

As before, the term on the right of inequality (3.31) is the predual function. If the weights are selected to cause the dependency of the predual function on the primal independent variables to disappear, the resultant expression is the dual function, $v(\bar{\delta})$

$$v(\bar{\delta}) = \left(\frac{c_1}{\delta_1}\right)^{\delta_1} \cdots \left(\frac{c_{m\,1}}{\delta_{m\,1}}\right)^{\delta_{m\,1}} \lambda_1^{\lambda_1}. \quad (3.34)$$

In order to achieve this reduction, it is required that

$$\sum_{i=1}^{m\,1} a_{ij}\delta_j = 0 \qquad j = 1, 2, \ldots, n. \quad (3.35)$$

Maximization of $v(\bar{\delta})$ subject to the orthogonality and normality conditions yields $v(\bar{\delta}*)$ which is also the minimum of the objective function satisfying the inequality constraint

$$\text{Min } z = v(\bar{\delta}*).$$

The degree of difficulty of a constrained geometric programming problem is $m_1 - n - 1$, which must be non-negative.

Solution (to Example 3.5) Since the objective function and constraint of example 3.5 are in the proper form, (3.34) may be applied directly to obtain the dual function

$$v(\bar{\delta}) = \left(\frac{2\pi}{\delta_1}\right)^{\delta_1} \left(\frac{10\pi}{\delta_2}\right)^{\delta_2} \left(\frac{1}{\pi\delta_3}\right)^{\delta_3} \lambda_1^{\lambda_1}$$

where
$$\delta_1 + \delta_2 = 1 ,$$
$$\delta_3 = \lambda_1$$

and $\qquad\qquad r: \quad \delta_1 + 2\delta_2 - 2\delta_3 = 0$

$\qquad\qquad\quad h: \quad \delta_1 - \delta_3 \qquad\quad = 0.$

Since there are four equations in four unknowns (δ_1, δ_2, δ_3, λ_1), a unique solution is obtained without maximizing $v(\bar{\delta})$; i.e., the problem has degree of difficulty zero.

The solution to these equations is

$$\delta_1 = 2/3 \quad \delta_2 = 1/3 \quad \delta_3 = \lambda_1 = 2/3.$$

Thus the cost of the sides of the vessel is twice the cost of the bottom plate (the same conclusion obtained in example 3.3), and, since $d_3 > 0$, the constraint is active; i.e., 10 trips are taken. The optimum value of r is obtained by the methods of section 3.2. Thus,

$$v(\bar{\delta}^*) = (3\pi)^{2/3} (30\pi)^{1/3} \left(\frac{3}{2\pi}\right)^{2/3} \left(\frac{2}{3}\right)^{2/3}$$

$$v(\bar{\delta}^*) = 9.42$$

and, since $10\pi(r^*)^2 = \frac{1}{3} (9.42)$,

$$r^* = 0.315 \text{ ft}; \quad h^* = 3.18 \text{ ft}.$$

Example 3.6 As director of research for the XYZ Company, it is your task to allocate this year's research budget of \$50,000 between two existing projects. The total budget, however, need not be completely allocated. Since each project is partially dependent upon the information gained in the other project, it is essential that both projects be supported. The predicted return from these two projects is

$$x_1^3 \ x_2^2$$

where x_i is the amount of money allocated to project i.

For reasons of morale, it is important that project 1 receive no more than three times the amount allocated to project 2. Determine the maximum return and the allocation policy which produces this return.

Solution Since geometric programming minimizes a posynomial, the objective function must be phrased as a minimization problem. Minimization of the negative of the return expression would not suffice, however, since the objective function would have a negative coefficient. However, minimizing the reciprocal of the return would; i.e.,

$$\text{Min } z = \frac{1}{x_1^3 x_2^2}$$

The constraints can then be written as

$$\text{morale:} \quad \frac{x_1}{3x_2} \leqslant 1$$

$$\text{budget:} \quad \frac{x_1}{50{,}000} + \frac{x_2}{50{,}000} \leqslant 1.$$

Note that there are a total of 4 terms in the objective function plus constraints and two independent variables. Hence, the problem has a degree of difficulty one.

The dual function is

$$v(\delta) = \left(\frac{1}{\delta_1}\right)^{\delta_1} \left(\frac{1}{3\delta_2}\right)^{\delta_2} \left(\frac{1}{50{,}000\,\delta_3}\right)^{\delta_3} \left(\frac{1}{50{,}000\,\delta_4}\right)^{\delta_4}$$

$$\cdot (\delta_2)^{\delta_2} (\delta_3 + \delta_4)^{\delta_3 + \delta_4}.$$

Normalization: $\delta_1 = 1$

Orthogonalization: $-3\delta_1 + \delta_2 + \delta_3 = 0$

$$-2\delta_1 - \delta_2 + \delta_4 = 0$$

We can express δ_2 and δ_4 in terms of δ_3:

$$\delta_2 = 3 - \delta_3$$

$$\delta_4 = 5 - \delta_3.$$

Hence, the dual function becomes, in terms of δ_3,

$$v(\overline{\delta}) = \left(\frac{1}{3}\right)^{3-\delta_3} \left(\frac{1}{\delta_3}\right)^{\delta_3} \left(\frac{1}{5-\delta_3}\right)^{(5-\delta_3)} \cdot$$

$$\cdot \left(\frac{1}{50,000}\right)^5 (5)^5.$$

A plot of $v(\overline{\delta})$ versus δ_3 is given on Figure 3.1. Observe that $v(\overline{\delta})$ has an unconstrained maximum at $\delta_3 = 15/4$.

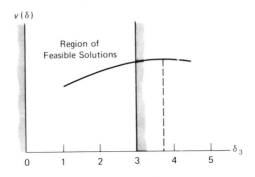

Figure 3.1

However, the weights are required to be non-negative. Since δ_2 becomes negative for $\delta_3 > 3$, the region of feasible solutions is $0 \le d_3 \le 3$. In this region, $v(\overline{\delta})$ is maximized at $\delta_3 = 3$. Hence

$$\delta_1 = 1$$

$$\delta_2 = 0 \quad \text{(constraint inactive)}$$

$$\delta_3 = 3$$

$$\delta_4 = 2.$$

The optimum allocation is x_1 = 3/5 of budget, x_2 = 2/5 of budget.

Bibliography

The definitive reference in this area is

1. Duffin, Richard J., Elmor L. Peterson and Clarence M. Zener. *Geometric Programming.* John Wiley & Sons, New York, 1967.

Other applications include:

2. Avriel, M. and D. J. Wilde. Optimal condenser design by geometric programming. *Ind. Eng. Chem. Process Des. Develop.*, 6, p. 256, 1967.

3. Kermode, R. I. Geometric programming: A simple efficient optimization technique. *Chem. Eng.*, , pp. 97—100, Dec. 1967.

The multistage compression example is presented in a classical thermodynamics background in:

4. Dodge, Barnett F. *Chemical Engineering Thermodynamics*, pp. 280—285. McGraw-Hill Book Co. New York, 1944.

4
Direct Optimization Techniques

4.1 INTRODUCTION

In chapter 1, it was pointed out that optimization techniques can be classified as direct or indirect, depending upon whether they use a comparison between function evaluations or whether they employ the necessary (Kuhn-Tucker) conditions. This distinction was not emphasized in the two preceding chapters, however, since our attention was confined to optimization systems whose structure permitted efficient solution algorithms to be developed. In problems of a more general nature, however, it is useful to distinguish between the two approaches and the techniques based on them.

Direct optimization techniques utilize evaluations of the objective function at specified values of the independent variable. Direct techniques must be able to recognize whether or not the maximum of the function occurs at one of the specified points. If further increase in the objective function is indicated, the technique must determine a new value of the independent variable where an additional function evaluation should be performed. This process can consist of discarding unproductive intervals of the independent variable to continue exploration in a promising interval. Alternatively, it may involve selecting a new extremum candidate based on observed trends in the objective function at prior evaluations. Tactics based on the former strategy are called elimination techniques; those based on the latter are referred to as climbing techniques.

Elimination techniques are well suited to single variable optimization problems and find ultimate expression in an optimum search algorithm. Unfortunately, they do not extend with comparable efficiency to multivariable problems and are seldom used in this regard.

Climbing techniques on the other hand are most effective in the multivariable context. The presentation will, therefore, be confined to elimination techniques for single variable optimization and climbing techniques for multivariable optimization.

4.2 SINGLE VARIABLE OPTIMIZATION

Direct optimization procedures can be introduced by considering the problem of maximizing a function f of a single independent variable x. Although it was shown in chapter 1 that the mathematical theory can be employed in principle to furnish the solution to such a problem, in practice it may be inconvenient or impossible to implement this theory. $f(x)$ may be a complicated function, and the obtainment of its derivative may involve considerable algebraic manipulation. Alternatively, $f(x)$ may be the result of an experiment or some other response generating procedure, precluding algebraic manipulation. In these situations it is more efficient to locate x^* by elimination techniques.

4.2.1 Elimination Techniques In order to illustrate the nature of elimination techniques, consider the selection of two points, x^1 and x^2, on the interval $[0,1]$ at which the objective function is to be evaluated. Suppose the results of this evaluation are as shown on Figure 4.1. On the basis of these results, can any finite interval of the region of feasible solutions — i.e., $[0,x^1]$, $[x^1,x^2]$, $[x^2,1]$, be discarded?

The answer to this question in general is no. Although the interval $[x^1,1]$ contains the larger of the two function values, no restriction was put on the behavior of $f(x)$. It is conceivable that the situation depicted in Figure 4.1 stemmed from an objective function such as depicted in

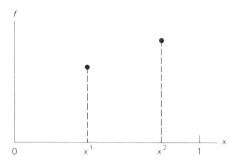

Figure 4.1

Figure 4.2, where x^* is to the left of x^1. For functions such as the one shown on Figure 4.2 where more than one relative maximum can occur, elimination techniques will have difficulty in locating the *global* optimum. For this reason, elimination techniques are best employed when the objective function has a single optimum point in the region of feasible solutions—more specifically when the function has the property of *unimodality*.

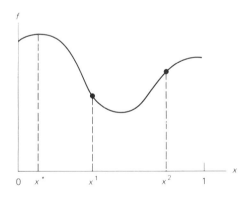

Figure 4.2

Definition 4.1 A function $f(x)$ is said to be unimodal on $[0,1]$ if it increases monotonically to its maximum, after which it decreases monotonically. That is, if

$$0 \leqslant x^1 < x^2 < x^* < x^3 < x^4 \leqslant 1,$$

Then

$$f(x^1) < f(x^2) < f(x*) > f(x^3) > f(x^4).$$

In other words, a unimodal function has a single hump. A concave function, for example, is unimodal. A function need not be concave nor even continuous to be unimodal, however.

If our attention had been restricted to unimodal functions in Figure 4.1, the optimum could not occur in the interval $[0, x^1]$. This interval could be discarded as far as further function evaluations are concerned.

Since the maximum value of $f(x)$ is now known to occur in the interval $[x^1,1]$—in which a function evaluation already has been made — we can repeat the elimination process by locating a third function evaluation point, x^3, in the interval $[x^1,1]$. Suppose the situation after evaluation of $f(x^3)$ is as depicted on Figure 4.3. By reasoning similar to that already used, the interval $[x^2,1]$ can now be eliminated.

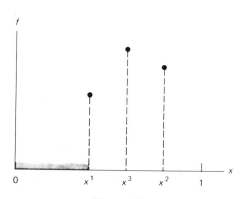

Figure 4.3

This procedure can be repeated in a sequential fashion until a solution of desired accuracy is obtained — that is, until the interval in which the maximum is known to occur is less than some preassigned resolution. If we consider the problem solved when $x*$ is known within $\pm\epsilon$, then ϵ is the *resolution* and the final interval will be 2ϵ in length. (This is the case, provided we can distinguish between function evaluations at x-values a distance ϵ apart. If it is impossible to determine which value of the function is larger for two values of x in a neighborhood of x^* larger than 2ϵ, then the final interval size is dictated by the *distinguishability* of the function.)

An elimination technique proceeding in this fashion is known as *sequential*, in that the function evaluations are performed one after the other and the information used successively to reduce the interval containing the extremum. This is in contrast to simultaneous elimination techniques in which all function evaluations are performed at the same time.

There may be occasions when simultaneous techniques are required. The interested reader is referred to several accounts of these methods in references (*1, 2*). Sequential techniques are considerably more efficient and should be used if at all possible.

A number of elimination procedures have been proposed for the sequential optimization of a unimodal function. These methods vary in the way in which points are selected for function evaluation. Consequently, they vary in efficiency. That is, for the same number of function evaluations, different methods bracket x^* in intervals of varying length. To put it another way, the number of function evaluations required to achieve a specified resolution varies from method to method. It is natural then to ask if there is a method which is most efficient. The answer is that there is. It is called Fibonacci Search.

4.2.2 Fibonacci Search In the thirteenth century, a man called Fibonacci mused on the properties of an integer series given by

$$F_0 = F_1 = 1$$

$$F_n = F_{n-1} + F_{n-2} \qquad n = 2, 3, \ldots \quad (4.1)$$

In 1953, a man named Kiefer showed that this series could be used to identify an optimum sequential elimination technique, which has come to be known as Fibonacci Search (*3*).

The nature of Fibonacci Search is easily illustrated by reference to Figure 4.4. N function evaluations are to be

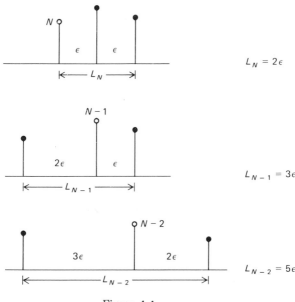

Figure 4.4

performed to provide a resolution ϵ. It is clear that the final three points must be located to form two equal intervals of length ϵ as shown in part a of Figure 4.4. The value of the function at the center point must be greater than the value of the function at either of the two outside points. (Note that, if the value of the function at the center point were equal to the value at one of the outside points, x^* would lie between the two points and L_N would equal ϵ. In general, one cannot presume the function will be so co-operative, and hence the argument must be based on the worst possible case.) The interval of uncertainty for bracketing x^* after the N function evaluations, L_N, is equal to 2ϵ.

Since these points were located sequentially, one must be the Nth point located. For illustration purposes, the lefthand point is arbitrarily selected on Figure 4.4, but the following argument holds regardless of which point was placed last.

The situation prevailing just prior to the Nth function evaluation is shown on part (b) of Figure 4.4. Comparison of parts (a) and (b) show that the Nth point was placed symmetrically in the L_{N-1} interval with respect to the already existing interior point. That is, the interior point is located a distance ϵ from the righthand boundary; hence the Nth point is located a distance ϵ from the lefthand boundary (and for the Nth step a distance ϵ from the other interior point). This symmetric location of succeeding points is a characteristic of Fibonacci Search, and it leads to the conclusion that the interval known to contain x^* after $N-1$ evaluations, L_{N-1}, is equal to 3ϵ.

The situation leading to part (b) of Figure 4.4 is now apparent, and one version is illustrated in part (c). The center point of L_{N-1}, is considered to have been the $(N-1)^{st}$ point located. Since it was placed symmetrically in the $(N-2)$ interval, L_{N-2} must equal 5ϵ. Continued application of this reasoning yields the following representation for the various interval lengths.

$$L_N \;=\; 2\epsilon \;=\; F_2\,\epsilon$$

$$L_{N-1} \;=\; 3\epsilon \;=\; F_3\,\epsilon$$

$$L_{N-2} \;=\; 5\epsilon \;=\; F_4\,\epsilon$$

$$\text{—} \text{—} \text{—} \text{—} \text{—} \text{—} \text{—} \text{—} \text{—}$$

$$L_2 \;=\; F_N\,\epsilon$$

$$L_1 \;=\; F_{N+1}\,\epsilon \tag{4.2}$$

where the F_k, $k = 2, 3, \ldots, N+1$ appearing in (4.2) are the Fibonacci numbers identified in (4.1).

Since the evaluation of the function at the first point reveals nothing about the location of x^*, L_1 must equal the original unit interval and the last equation in (4.2) can be written as

$$\epsilon \;=\; \frac{1}{F_{N+1}} \tag{4.3}$$

Equation (4.3) can be used either for design or rating purposes. If a resolution ϵ is given, the number of evaluations required may be obtained from the index of the largest Fibonacci number whose reciprocal is greater than or equal to ϵ. Conversely, if N evaluations are planned, a resolution no greater than $1/F_{N+1}$ can be expected.

The implementation of Fibonnaci Search is quite simple. Two initial function evaluations are performed at $x^2 = L_2 = F_N \epsilon$ and $x^1 = 1 - L_2 = 1 - F_N \epsilon$. Depending upon the outcome, the interval $[0, x^1]$ or $[x^2, 1]$ is eliminated and x^3 is located in the remaining interval, symmetrically with respect to the interior point. The process is repeated until all N function evaluations have been performed.

Example 4.1 Determine to within $\pm\ 0.05$ the value of x on the unit interval which maximizes the function $x(1.5 - x)$.

Solution For $\epsilon = 0.05$, by (4.2)

$$\frac{1}{F_{N+1}} = 0.05.$$

Since $F_7 = 21$, we see that 6 evaluations are required. $F_6 = 13$. Hence,

$$x^1 = 1 - 13(0.5) = 0.35;\ f(x^1) = 0.4025$$

$$x^2 = \quad\quad 13(0.5) = 0.65;\ f(x^2) = 0.5525.$$

Since $f(x^2) > f(x^1)$, the interval $[0, x^1]$ is eliminated (Figure 4.5). x^3 is located a distance $(x^2 - x^1) = 0.3$ from the right hand point; $x^3 = 0.7$;

$$f(x^3) = 0.56.$$

Since $f(x^3) > f(x^2)$, the interval $[x^1, x^2]$ is eliminated and x^4 is 0.95;

$$f(x^4) = 0.5225.$$

Since $f(x^4) < f(x^3)$, the interval $[x^4, 1]$ is eliminated and x^5 is 0.90;

$f(x^5) = 0.54$.

Since $f(x^5) < f(x^3)$, the interval $[x^5, x^4]$ is eliminated and x^6 is 0.85;

$f(x^6) = 0.5525$.

Since $f(x^6) < f(x^3)$, the interval $[x^6, x^5]$ is eliminated and $x^7 = 0.80$;

$f(x^7) = 0.56$.

Since $f(x^7) = f(x^3)$, the intervals $[x^7, x^6]$ and $[x^2, x^3]$ are eliminated, and we conclude

$0.7 < x^* < 0.8$.

The correct answer is $x^* = 0.75$.

4.2.3 Near-Optimum Sequential Search

Although Fibonacci Search is the most efficient sequential elimination technique, there is occasionally a reluctance to employ it because of the necessity to specify beforehand either the number of function evaluations, N, or the resolution, ϵ. Fortunately a near-optimum alternative results from the application of a limiting argument to the Fibonacci sequence.

It is possible to rewrite the formula for L_2 by combining the L_2 expression in (4.2) with the expression for ϵ from equation (4.3):

$$L_2 = \frac{F_N}{F_{N+1}} \qquad (4.4)$$

Suppose L_2 is selected on the premise that a very large number of function evaluations will be made (even though we may have no intention of performing them). Call the approximation for L_2 so obtained $\bar{L}_2$:

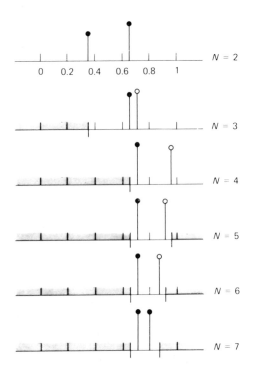

Figure 4.5

$$\bar{L}_2 = \lim_{N \to \infty} \frac{F_N}{F_{N+1}}. \qquad (4.5)$$

If this limit exists, then for large N

$$\bar{L}_2 = \frac{F_N}{F_{N+1}} = \frac{F_{N+1}}{F_{N+2}}.$$

Replacement of F_{N+2} by its definition (4.1) yields

$$\bar{L}_2 = \frac{F_N}{F_{N+1}} = \frac{F_{N+1}}{F_{N+1} + F_N} \qquad (4.6)$$

or $$\bar{L}_2 = \frac{F_N}{F_{N+1}} = \frac{1}{1 + F_N/F_{N+1}} = \frac{1}{1 + \bar{L}_2}. \quad (4.7)$$

Rearrangement of (4.7) gives

$$\bar{L}_2 + (\bar{L}_2)^2 = 1 \quad (4.8)$$

whose solution is

$$\bar{L}_2 = \frac{-1 \pm \sqrt{5}}{2}.$$

Since $\bar{L}_2$ must be positive,

$$\bar{L}_2 = \frac{\sqrt{5} - 1}{2} = 0.618. \quad (4.9)$$

This value of $\bar{L}_2$ provides for the placement of the initial points x^1 and x^2 at 0.382 and 0.618, respectively, independent of N or ϵ. The solution algorithm then proceeds in exactly the same manner as previously discussed.

It is easy to show that, for this initial location, the ratio of any two succeeding intervals is a constant, equal to 0.618. This ratio was known to ancient architects and geometricians as the golden section. It divides a line segment into two parts such that the ratio of the larger to the original segment equals the ratio of the smaller to the larger. For this reason, this elimination technique is sometimes referred to as search by golden section.

4.2.4 Hybrid Elimination Methods A distinguishing feature of the elimination procedures just described is the fact that, once an initial function value is obtained, intervals are eliminated as a result of a single additional function evaluation. It is possible to combine a sequential and simultaneous approach by performing the function evaluations in a sequential series of blocks, each block containing a specified number of simultaneous evaluations. This hybrid technique requires a larger number of function

evaluations to obtain the same resolution as obtained with
an exclusively sequential technique. However, the number
of blocks required is less than the number of single evaluations, and the technique could have merit when the evaluation procedure involves long delay times — e.g., in exper-
imental situations.

The concept of odd-block search was developed by
Douglass Wilde and his students. It is discussed in detail in
the text by Wilde and Beightler (2).

4.3 MULTIVARIABLE OPTIMIZATION

In this section, we return to the general parameter
optimization problem introduced in chapter 1 — namely
to find the non-negative values of n independent variables, $x_1, x_2, \ldots, x_n$ which maximize a given function of these variables

$$z = f(x_1, x_2, \ldots, x_n) \qquad (4.10)$$

while satisfying the following constraint equations

$$g_i (x_1, x_2, \ldots, x_n) = 0 \qquad i = 1, 2, \ldots, m \qquad (4.11)$$

$$x_j \geqslant 0 \qquad j = 1, 2, \ldots, n \qquad (4.12)$$

Although it is possible for direct optimization methods
to handle constraint equations such as (4.11) and (4.12),
we first consider the unconstrained case.

Geometrically, the objective function constitutes an n-
dimensional surface in the $n + 1$ dimensional space consisting of the independent variables, $x_1, x_2, \ldots, x_n$
and the dependent variable, z. Representation of such a
surface is difficult on a two-dimensional printed page.
Consequently, it is often convenient to resort to a *contour plot* of the function, in which the value of z for a
given $\bar{x}$ is projected onto the $\bar{x}$ plane. Equal values of z
are then connected by *contour lines* in the same manner as
a topographic map.

For example, consider the function of two variables

$$z = 10 - 2(x_1 - 1)^2 - (x_2 - 1)^2.$$

A contour plot of this function is shown in Figure 4.6.

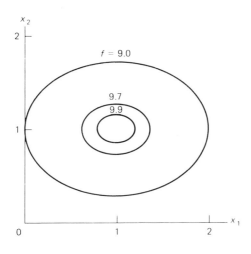

Figure 4.6

The contour lines are ellipses with centers at $x_1 = 1, x_2 = 1$, which is also the maximizing point.

Direct multivariable optimization techniques can be depicted as a progression of test points on the $x_1 x_2$ plane. At each point, a function evaluation is performed and a decision made to continue the search for the optimum or to terminate the search. Unlike linear programming, the test points are not vertices of a convex region.

4.3.1 Sectioning The simplest progression of test points is one in which only a single variable is changed at a time. The other $n - 1$ variables are held constant. The objective function is thus reduced to a function only of this variable and either analytical or one-dimensional search procedures can be used to find its maximum. The variable is assigned the value providing this maximum and one of the remaining variables is selected as the next search candidate. This procedure is repeated until all n of the independent variables have been examined sequentially. The entire process is then repeated until no further increase in the objective function is observed. This technique is known as *sectioning*.

Example 4.2 Obtain by sectioning the values of x_1 and x_2 which maximize the function.

$$z = 10 - 2(1 - x_1)^2 - (1 - x_2)^2.$$

Solution The first step in solving optimization problems by climbing techniques is to select an initial point. This selection is arbitrary, but, if any intuition or a promising starting point is available, it should generally be used. We arbitrarily start at the origin and choose to keep x_2 fixed and search first on x_1. With $x_2 = 0$,

$$z = 9 - 2(1 - x_1)^2,$$

and it is obvious that z is maximized for $x_1 = 1$.

We next set $x_1 = 1$ and search on x_2. With $x_1 = 1$,

$$z = 10 - (1 - x_2)^2;$$

again, it is obvious that $x_2 = 1$ maximizes z. Since we have finished our initial exploration, we must repeat the process. Thus, we set $x_2 = 1$ and search on x_1

$$z = 10 - 2(1 - x_1)^2.$$

Since $x_1 = 1$ is again the maximizing value, we correctly conclude the optimum has been found:

$$x_1^* = 1, \; x_2^* = 1, \; z = 10.$$

Sectioning was effective in solving the problem in example 4.2 because the objective function had no interactions — i.e., $x_1 x_2$ cross products. When an objective

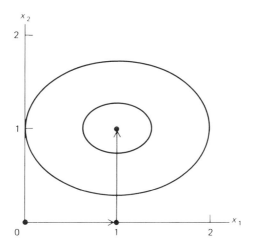

Figure 4.7

has such terms, sectioning can be inefficient or even ineffective in locating the maximum. To illustrate this, consider the contour plot of part (a), Figure 4.8. Here there are weak interactions, creating a function surface with a mild ridge. In order to ascend this ridge, the sectioning search path zig-zags back and forth, requiring many function evaluations. If the ridge sharpens due to strong interactions, as illustrated by part (b) of Figure 4.8, it is highly probable that the sectioning search path will reach a point on the ridge line. Finite exploration in directions parallel to the x_1 and x_2 axes will indicate that no further function increase is possible. The sectioning algorithm will incorrectly identify the point as the optimum one, whereas a simple glance up the ridge will reveal this is not so.

In order to avoid having the search path become trapped on a ridge line, a number of optimization procedures have been developed in which the search path is permitted to follow a direction other than parallel to the coordinate axes. This increased flexibility considerably improves the efficiency of direct search methods without introducing undue complications.

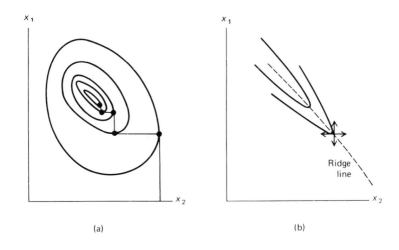

(a)   (b)

Figure 4.8

4.3.2 Pattern Search In 1961, R. Hooke and T.A. Jeeves of the Westinghouse Corporation described a direct search technique which has the ability to ascend straight ridges. The technique is known as *pattern search* (7).

We can illustrate pattern search by reference to Figure 4.9. Pattern search begins with a *local exploration* about a selected *original base point* $\bar{x}^1$. This local exploration consists of changing a single independent variable (x_i) at a time by a preset amount or step size (d_i) which can be individually tailored for each variable. The exploration evaluates the function at its original base point and at perturbations in the search variable about this point of $\pm d_i$. If one of the changes results in an increase of the objective function, we assign the search variable this new value and consider the point as a *temporary base point*, $\bar{x}_t$. If neither change produced a *success* (increase in function value), we reduce the step size for the variable and repeat the procedure. Unless we have begun our search at the maximum point or on a ridge, this exploratory search will locate a new temporary base point.

On Figure 4.9, we show an exploratory search being carried out first on x_1. The perturbation $+ d_1$ failed to yield improvement, but $- d_1$ did. Hence, a temporary

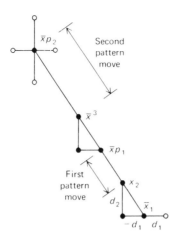

Figure 4.9

base point is established.

Another independent variable is selected, and explora-tory search is carried out with respect to this variable about the temporary base point. If this exploratory search is successful, a new temporary base point is established to replace the old one. The process is repeated sequentially until all the independent variables have been explored. The final temporary base point is identified as the second base point, $\bar{x}^2$.

On Figure 4.9, we have indicated a success for a $+ d_2$ perturbation about the first temporary base point. Since there are only two independent variables, this success iden-tifies the second base point, $\bar{x}^2$.

The original and second base points create a pattern which is used to locate the first *pattern point*, $\bar{x}_{p1}$; $\bar{x}_{p1}$ is the terminal point of a directed line segment from $\bar{x}^1$ through $\bar{x}^2$, twice the length of the line between $\bar{x}^1$ and $\bar{x}^2$. By this location of $\bar{x}_{p1}$, it is assumed that an explor-atory search about $\bar{x}^2$ will meet with similar success as about $\bar{x}^1$.

The function is first evaluated at $\bar{x}_{p1}$ to determine if $f(\bar{x}_{p1}) > f(\bar{x}^2)$. If it is, the pattern move has been suc-cesful, and local exploration is made about $\bar{x}_{p1}$. This exploration will result in a third base point, $\bar{x}^3$.

The local exploration about $\bar{x}_{p1}$ leading to $\bar{x}^3$ is shown on Figure 4.9.

The base points $\bar{x}^2$ and $\bar{x}^3$ are used to make a pattern move, which locates the second pattern point $\bar{x}_{p2}$. Notice that this pattern move is twice the length of the first. If continued exploration were to reveal similar successes, each succeeding pattern move would be larger than its predecessor. Thus, pattern search provides an effective *acceleration* in the step sizes used in the search routine. This is an important characteristic of the method. It insures a rapid ascent up a straight ridge, even if the original step sizes were small.

When a point is reached where local exploration reveals no improvement, the search is either at the maximum or at a point where the ridge is turning. The tactic then is to reduce the step sizes of the individual variables and repeat the local exploration. Reduction of the step sizes is continued until they are below the specified resolutions for the variables or until a function improvement is obtained. The former case means the desired maximum has been found. The latter case means the pattern search procedure must restart in the indicated direction.

Example 4.3 Maximize the following objective function by pattern search

$$z = 1 - 2(1 - x_1)^2 - (1 - x_2)^2 - (2x_1 + x_2 - 1)^2.$$

Solution The contour plot of this function is given in Figure 4.10. Let us select

$$d_1 = 0.1 \qquad d_2 = 0.1$$

and consider our search completed if both the step sizes should decrease below 0.01.

Initiate the search at the point $(1,0)$. Here $z = -1$. Perturb x_1 first: for $(1.1,0)$ $z = -2(.1)^2 - (1.2)^2 = -1.46$.

for $(0.9,0)$ $z = -2(.1)^2 - (0.8)^2 = -0.66$.

Hence, the point $(0.9,0)$ is a temporary base point.

Next, perturb x_2 for (0.9, 0.1) z = 1 $-2(.1)^2 - (0.9)^2$ = -0.64.

This move is successful. The point (0.9, 0.1) is thus the second base point, and a pattern move is in order. This move locates the first pattern point (0.8, 0.2) where

$$z = 1 - 2(0.2)^2 - (0.8)^2 - (0.8)^2 = -0.36.$$

The pattern move was successful.

Local exploration begins anew about the pattern point. First an x_1 perturbation
for (0.7, 0.2), z = 1 $-2(0.3)^2$ $-(0.8)^2$ $-(0.6)^2$ = -0.18.
This move is successful and defines a temporary base point at (0.7, 0.2). Perturbation of x_2 gives
for (0.7, 0.3) z = 1 $-2(0.3)^2 - (0.7)^2 - (0.7)^2$ = -0.16.
Hence the point (0.7, 0.3) is the third base point. A new pattern point is now obtained by projecting a line from (0.9, 0.1) through (0.7, 0.3). This locates (0.5, 0.5) as the second pattern point
for (0.5, 0.5) z = 1 $-2(0.5)^2 - (0.5)^2 - (0.5)^2$ = 0.

Thus the acceleration procedure is successful, and local exploration begins anew about (0.5, 0.5).

As it turns out, the point (0.5, 0.5) is the optimum as local exploration reveals. Continued decrease in the step size produces no further increase in the objective function.

It must be obvious to the reader that the direction of search plays a vital role in the success of the pattern search technique. It might be asked if there is a formal procedure for determining a preferred direction of search. The answer is that there is, and many optimization techniques have been developed which specifically compute a preferred search direction. These techniques involve the evaluation of the partial derivatives of the objective function and are referred to as *gradient techniques.* Because they invoke the necessary conditions for an extremum, they are discussed in the next chapter. However, it should be noted here that such techniques are generally more sophisticated and more difficult to understand than pattern search. Because of its simplicity, pattern search is easily programmed for a digital computer. Modifications, when necessary, can be carried out with little difficulty.

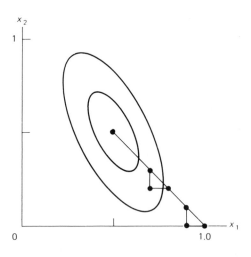

Figure 4.10

Many comparative evaluations of optimizing procedures have been carried out. Unfortunately, these evaluations rarely agree, and it is impossible to state that any single technique is superior in all cases. However, there is continuing evidence that pattern search does a satisfactory job in solving a variety of engineering problems.

4.3.3 Handling Constraints There is no easy provision for handling constraints by direct optimizing techniques. However, many users have been able to solve constrained problems with direct methods by use of *penalty functions.*

The constrained optimization problem posed by (4.10) and (4.11) is transformed into an unconstrained one in terms of a modified objective function Z

$$Z = z - \sum_{i=1}^{m} p_i \, g_i^2 \qquad (4.13)$$

where the p_i are large positive constants. The terms $p_i g_i^2$ are called *penalty functions.*

As the p_i increase, it can be shown (11) that the solution to the problem posed by (4.13) converges to the original problem. In actual practice, the choice of p_i is critical. The recommended procedure is to solve the problem several times, progressively increasing the p_i.

Example 4.4

$$\text{Maximize } z = 1 - 2(1 - x_1)^2 - (1 - x_2)^2$$

$$\text{subject to } 2x_1 + x_2 - 1 = 0.$$

This is the objective function of example 4.2 except that an equality constraint has been added.

Solution It is easily shown that

$$x_1^* = 1/3, \; x_2^* = 1/3.$$

Let us use the penalty function concept to solve the problem

$$Z = 1 - 2(1 - x_1)^2 - (1 - x_2)^2 - P(2x_1 + x_2 - 1)^2.$$

Analytically,

$$\frac{\partial Z}{\partial x_1} = 4(1 - x_1) - 4P(2x_1 + x_2 - 1) \overset{set}{=} 0$$

$$\frac{\partial Z}{\partial x_2} = 2(1 - x_2) - 2P(2x_1 + x_2 - 1) \overset{set}{=} 0.$$

Since these equations are linear, we obtain their solution by Cramer's Rule

$$x_1 = \frac{1 + \dfrac{1}{P}}{3 + \dfrac{1}{P}}$$

$$x_2 = \frac{1 + \dfrac{1}{P}}{3 + \dfrac{1}{P}}$$

Thus, as $P \to \infty$ $x_1 \to x_1^*$ and $x_2 \to x_2^*$.

By Direct Optimization Techniques Rather than attempt to solve this problem by a direct technique, we instead plot the contours of Z for $P = 1$ and $P = 100$ in Figure 4.11. Notice the sharp ridge for $P = 100$. If we were to utilize sectioning to solve this problem with $P = 100$, we would undoubtedly become trapped somewhere on the ridge line. Pattern search would probably converge to the proper solution, although it could require several function evaluations.

Alternatively, we could begin by locating the extremum for $P = 1$. This is then the objective function of example 4.3 and pattern search will successfully locate the solution $(1/2, 1/2)$. If we retain this point as the starting point for a search on Z with $P = 5$, we should be able to locate the point $(3/8, 3/8)$. Again we use this point as the initial point to locate the maximum of Z for $P = 10$. This would locate the point $(11/31, 11/31)$. Further repetition should locate $(1/3, 1/3)$ to the desired accuracy.

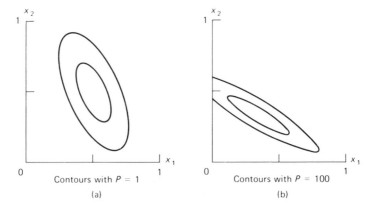

Figure 4.11

Bibliography

4.2 SINGLE VARIABLE OPTIMIZATION

Additional elimination techniques are discussed in

1. Wilde, Douglass J. *Optimum Seeking Methods*. Prentice-Hall, Inc., Englewood Cliffs, New Jersey, 1964.

2. Wilde, Douglass J. and Charles S. Beightler. *Foundations of Optimization*. Prentice-Hall, Inc. Englewood Cliffs, New Jersey, 1967.

Fibonacci search was first proposed by Kiefer:

3. Kiefer, J. Sequential minimax search for a maximum. *Proc. Am. Math. Soc.*, 4, pp. 502—506, 1953.

4. Kiefer, J. Optimum sequential search and approximation methods under minimum regularity assumptions. *J. Soc. Ind. Appl. Math.*, 5, p. 125, 1957.

Other proofs of the optimality of Fibonacci search have been given—e.g.,

5. Johnson, S.M. *Optimal Search for a Maximum is Fibonaccian*. RAND Corp. Report p. 856, 1956.

An interesting variant of Fibonacci search is provided if a cost is assigned to each experiment. Then the search procedure is designed to minimize a weighted sum of the final interval length and the cost of conducting the search. This idea is discussed in

6. Eichhorn, B.H. On Sequential Minimax. *J. Math. Anal. Appl.*, 14, pp. 31—37, 1966.

4.3 MULTIVARIABLE OPTIMIZATION

Pattern Search was proposed in 1961:

7. Hooke, R. and T.A. Jeeves. Direct search solution of numerical and statistical problems. *J. Ass. Comput. Mach.*, 8, pp. 212—229, 1961.

Some applications and extensions of the method are given in

8. Weisman, J. and C.F. Wood. The use of optimal search for engineering design. *Recent Advances in Optimization Techniques. (Abrahim Lavi and Thomas P. Vogl, eds.)*. John Wiley & Sons, Inc., New York, 1966.

9. Weisman, Joel, C.F. Wood and L. Rivlin. Optimal design of chemical process systems, *C.E.P. Symposium Series*, 61, pp. 50—63, 1965.

An attempt to evaluate optimization routines in the design context is given by

10. Wood, Cyrus F. Review of design optimization techniques, *IEEE Transactions on Systems Science and Cybernetics*, 1, pp. 14—20, 1965.

Proof of the penalty function convergence is found in

11. Fiacco, A.V. and G.P. McCormick. Extension of sequential unconstrained minimization techniques for non-linear programming: equality constraints and extrapolation. *Manage. Sci.*, 12, pp. 816—828, 1966.

5
Indirect Optimization Techniques

5.1 INTRODUCTION

By and large, most of the techniques proposed for the solution of the general parameter optimization problem are indirect techniques. They seek, by some process or other, the point which satisfies the necessary conditions of the problem. Over the years these techniques have come to be known by many names: hill climbing routines, gradient methods, nonlinear programming, etc.

It would be impossible to discuss or even mention all of the techniques which have been proposed for solving the general parameter optimization problem. Moreover, not all of the proposed methods are equally effective in achieving their goal. Rather, the approach taken in this chapter is to introduce the more important concepts of indirect optimization and to demonstrate how these concepts have been used to construct a number of the more successful optimization routines.

Once again we are seeking to maximize a function of n independent variables

$$\text{Max } f(\bar{x})$$

$$\text{subject to } g_i(\bar{x}) = 0 \qquad i = 1, 2, \ldots, m$$

$$x_j \geqslant 0 \qquad j = 1, 2, \ldots, n.$$

We will assume f has continuous first partial derivatives and the constraints provide a convex region of feasible solutions.

5.2 GRADIENT METHODS

All effective multivariable optimization techniques utilize the derivatives of the objective function at a test point. The use of these derivatives varies somewhat from technique to technique. In this section, we will introduce methods which use only the derivatives at the current test point to provide a search direction.

First, a few definitions are in order.

Definition 5.1 A *contour tangent* of $f(\bar{x})$ at the point $\bar{x}^k$ is the tangent to the contour at this point. Since f is constant along a contour, the change in f, Δf is zero:

$$\Delta f = 0.$$

The contour tangent is defined by the equation

$$\Delta f = \sum_{i=1}^{n} \left(\frac{\partial f}{\partial x_i} \right)_k (x_i - x_i^k) = 0 \qquad (5.1)$$

where $\left(\dfrac{\partial f}{\partial x_i} \right)_k$ denotes the partial derivative of f with respect to x_i — evaluated at $\bar{x}^k$. Figure 5.1 illustrates a contour tangent for a two dimensional situation.

Definition 5.2 A direction in the space of n independent variables is defined by a set of n values, $\bar{r} = (r_1, r_2, \ldots, r_n)$, the direction being that of the directed line from the origin to $\bar{r}$. For convenience we assume $\bar{r}$ to be normalized; i.e.,

$$\sum_{i=1}^{n} r_i^2 = 1.$$

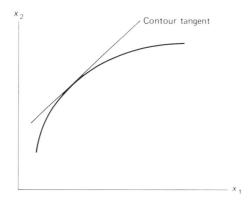

Figure 5.1

In 2-dimensional space, the direction $(1/\sqrt{2}, 1/\sqrt{2})$ would be the $45°$ line directed from the origin to the point $(1/\sqrt{2}, 1/\sqrt{2})$.

The concept of direction is useful in defining another quantity. The partial derivatives of a function indicate the rate of change of a function in the direction of the co-ordinate axes. One can define a derivative of f in any direction desired. Such a derivative is called a *directional derivative*.

Definition 5.3 If all the first partial derivatives of the function f are continuous, the directional derivative of f in the direction $\bar{r}$, evaluated at the point $\bar{x}^k$ is written as $D_r\ f(\bar{x}^k)$ and given by

$$D_r\ f(\bar{x}^k) = \sum_{i=1}^{n} \left(\frac{\partial f}{\partial x_i} \right)_k r_i \qquad (5.2)$$

5.2.1 Steepest Ascent We now use these definitions to develop a basis for gradient methods. For the moment, consider only the unconstrained maximization of the function f, and assume that it is possible to evaluate the partial derivatives of f at any point.

Consider that, as a result of previous maneuvers, the current search point is $\bar{x}^k$. Consider further that all the partial derivatives of f at $\bar{x}^k$ have been computed and the Kuhn-Tucker conditions are not satisfied — i.e.,

$$x_i^k \left(\frac{\partial f}{\partial x_i} \right)_k \neq 0.$$

The question at this point is how to proceed. A plausible way is to specify a direction, $\bar{r}$, and seek the maximum of f proceeding in this direction from $\bar{x}^k$. This reduces the problem to a single variable optimization problem in terms of d the distance along the line from $\bar{x}^k$ in the direction $\bar{r}$. This one-dimensional problem can then be solved by one of the analytical or search techniques already discussed.

One way to specify $\bar{r}$ is to require that it point in the direction in which the function increases most rapidly. That is, we select $\bar{r}$ such that the directional derivative of f is maximized at $\bar{x}^k$:

$$\underset{\{r_i\}}{\text{Max}} \ D_r \ f(\bar{x}^k) = \sum_{i=1}^{n} \left(\frac{\partial f}{\partial x_i} \right)_k r_i$$

$$\text{subject to} \ \sum_{i=1}^{N} r_i^2 = 1.$$

This problem can be solved with the aid of Lagrange multipliers. The Lagrangian function is

$$F = \sum_{i=1}^{n} \left(\frac{\partial f}{\partial x_i} \right)_k r_i + \lambda \left(1 - \sum_{i=1}^{n} r_i^2 \right).$$

The requirement that F be made stationary implies

$$\frac{\partial F}{\partial r_j} = \left(\frac{\partial f}{\partial x_j} \right)_k - 2\lambda r_j = 0 \qquad j = 1, 2, \ldots, n$$

or

$$r_j = \frac{1}{2\lambda} \left(\frac{\partial f}{\partial x_j} \right)_k. \qquad (5.3)$$

By squaring (5.3) and summing over j, we obtain

$$\sum_{j=1}^{n} r_j^2 = 1 = \frac{1}{4\lambda^2} \sum_{j=1}^{n} \left(\frac{\partial f}{\partial x_j} \right)_k^2.$$

Hence

$$\lambda = \pm \frac{1}{2} \sqrt{\sum_{j=1}^{n} \left(\frac{\partial f}{\partial x_j} \right)_k^2}.$$

and

$$r_j^* = \pm \frac{\left(\dfrac{\partial f}{\partial x_j} \right)_k}{\sqrt{\sum_{j=1}^{n} \left(\dfrac{\partial f}{\partial x_j} \right)_k^2}} \qquad j = 1, 2, \ldots, n. \qquad (5.4)$$

Each component of the optimum direction is equal to plus or minus the corresponding normalized partial derivative of the function. The plus sign indicates the direction of maximum function increase; the minus sign indicates maximum function decrease.

The individual partial derivatives $\left(\dfrac{\partial f}{\partial x_1}, \ldots, \dfrac{\partial f}{\partial x_n} \right)$ themselves constitute a direction in the space of independent variables. This direction is called the *gradient direction*. We have just shown that a function increases most rapidly in its gradient direction.

Since the partial derivatives of f with respect to the independent variables must be computed at a test point to see if the Kuhn-Tucker conditions are satisfied, it is a simple matter to specify the gradient direction for continued search.

Geometrically, it is easy to identify the gradient direction by means of the contour tangent, since the gradient direction is normal to the contour tangent. To demonstrate that this is so, consider the function

$$f = 1 - x_1^2 - x_2^2.$$

The contour for $f = -1$ is a circle with center at the origin and radius equal to $\sqrt{2}$. The partial derivatives of f at the point (1, 1) on the contour are

$$\frac{\partial f}{\partial x_1} = -2x_1 = -2$$

$$\frac{\partial f}{\partial x_2} = -2x_2 = -2.$$

Hence, $r_1^* = \pm \frac{1}{\sqrt{2}}$, $r_2^* = \pm \frac{1}{\sqrt{2}}$

and the gradient direction is the 45° line.

The equation for the contour tangent is

$$(-2)(x_1 - 1) + (-2)(x_2 - 1) = 0$$

or $2x_1 + 2x_2 = 4.$

Since the slope of this line is −1, it is obviously perpen-

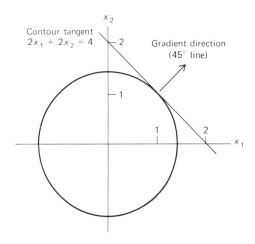

Figure 5.2

dicular to the gradient direction.

We can use this geometric property to trace the progress of a gradient technique without actually solving the problem. Consider that a two-dimensional objective function has a contour plot as shown in part (a) of Figure 5.3. Because the contours are nearly circles, the direction of steepest ascent locates the maximum very quickly. In part (b) the contours are elongated ellipses, creating a sizable ridge. If the gradient search begins at the foot of this ridge, its ascent is zig-zagged and inefficient.

Herein lies one of the unfortunate aspects of the steepest ascent technique. The contour plot shown in part (b) of Figure 5.3 is the same function shown in part (a) except that the variable x_1 has undergone a scale change. This simple distortion has resulted in an inefficient gradient search path. Thus, the speed of convergence of a gradient technique depends upon the choice of scales for the independent variables. This observation leads to a simple rule of thumb for choosing the scales of measurement for the independent variables. The scales should be selected to make the contours as spherical as possible. If the contours are spherical, the gradient direction from any base point passes through the optimum and the search is completed in one step. A corrolary of this rule is to avoid function representations involving transcendental terms (such as exponentials) if at all possible. Such representations create sharp ridges in the function surface which slow the progress of gradient techniques.

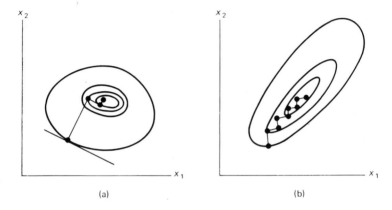

(a)   (b)

Figure 5.3

Unfortunately it is not always possible to apply these rules in framing an optimization problem, particularly if the function is complex. However, they can serve as guidelines, particularly in troubleshooting a slowly convergent problem.

5.2.2 Linear Inequality Constraints In this section, we consider a form of the general parameter optimization problem in which the constraints are linear inequalities:

$$\text{Max } f(x_1, x_2, \ldots, x_n) \tag{5.5}$$

$$\text{S.T. } \sum_{j=1}^{n} a_{ij} x_j \leqslant b_i \qquad i = 1, 2, \ldots, m. \tag{5.6}$$

$$x_j \geqslant 0 \tag{5.7}$$

It is not desirable to convert the inequalities to equalities by addition of slack variables. However, as in the case of linear programming, we assume the region of feasible solutions defined by (5.6) and (5.7) to be convex.

Linear equality constraints are not considered since they could be used to eliminate the equivalent number of variables, leaving the problem as stated.

As in the unconstrained case, the solution to this problem will be obtained by following a directed path through the region of feasible solutions rather than moving from vertex to adjacent vertex as in linear programming. Thus we initiate the solution by selecting an initial feasible point, $\bar{x}^0$. Unlike linear programming, this point need not correspond to a basic feasible solution.

In order to determine if the point $\bar{x}^0$ satisfies the Kuhn-Tucker conditions, the partial derivatives of f with respect to the independent variables are calculated. Since it would be fortuitous indeed if $\bar{x}^0$ were a local maximum, it can be safely assumed that not all the partial derivatives are zero and hence a direction of steepest ascent exists. Unlike the unconstrained case, this direction may or may not point within the region of feasible solutions. Hence, we must be prepared for two eventualities:

(a) the direction of steepest ascent points within the region of feasible solutions;
(b) it does not.

These are illustrated in Figure 5.4.

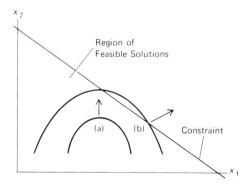

Figure 5.4

Case (a) is considered first since it is similar to the situation encountered in unconstrained steepest ascent. The search sets off in the gradient direction and is a single variable optimization problem in terms of the distance in this direction from $\bar{x}^0$, d. Unlike the unconstrained case, there is a restriction on d in that it cannot locate a point outside of the region of feasible solutions. If we call d_{Max} the maximum distance, one can proceed in the gradient direction without violating one of the constraints; then the search in the gradient direction can be expressed mathematically as

$$\underset{0 < d \leqslant d_{\text{Max}}}{\text{Max}} \; f(x_1^0 + \left(\frac{\partial f}{\partial x_1}\right)_0 d, \; \ldots \; , \; x_n^0 + \left(\frac{\partial f}{\partial x_n}\right)_0 d).$$

$$(5.8)$$

Equation (5.8) is to be read as follows: find the value of d, $0 < d \leqslant d_{Max}$ which maximizes f. This maximization may be carried out analytically or by means of one of the elimination techniques described in chapter 4.

Case (b) presents a different problem. Because the steepest ascent direction leads outside of the feasible region, an alternate direction must be selected. One way to select this direction would be to maximize the directional derivative subject to the constraint of remaining within the region of feasible solutions. For the case of linear constraints, a simpler alternative exists.

Figure 5.5 depicts the situation. The direction $\bar{r}$ forms an angle θ with the gradient direction. The smallest non-obtuse angle satisfying the constraints identifies a direction in which the objective function must increase.

Another way of looking at the situation is obtained by *projecting* the gradient direction onto the constraint. This means that a line is drawn from the arrowhead depicting the gradient direction, perpendicular to the constraint. The direction from $\bar{x}^0$ to the intersection of the perpendicular line and constraint is the direction which provides constrained ascent.

As in part (a) the distance from $\bar{x}^0$ in the direction so obtained must be selected to maximize f subject to the condition that no constraints be violated. Equation (5.8)

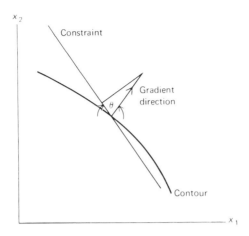

Figure 5.5

would again apply.
  Example 5.1

Maximize $f = (x_1 - 1)^2 + x_2$

Subject to $x_2 \leqslant 2$ (i)

$6x_1 + x_2 \leqslant 16$ (ii)

$x_1 \geqslant 0 \;\; x_2 \geqslant 0$

Solution The contours and region of feasible solutions
for this problem are shown in Figure 5.6A, part (a).
  Part (b) of Figure 5.6A traces out the following solution.
Consider $\bar{x}^0 = (2, 1/2)$. The partial derivative values at
$\bar{x}^0$ are

$$\left(\frac{\partial f}{\partial x_1} \right)_0 = 2(x_1^0 - 1) = 2(2 - 1) = 2$$

$$\left(\frac{\partial f}{\partial x_2} \right)_0 = 1.$$

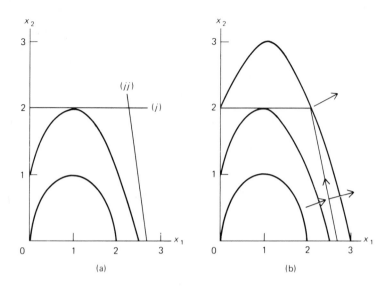

Figure 5.6A

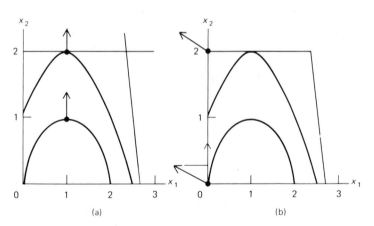

Figure 5.6B

The partial derivatives are not zero, so the gradient direction is $\left(\dfrac{2}{\sqrt{5}}, \dfrac{1}{\sqrt{5}}\right)$. Since this is a feasible direction, we must locate the distance d which maximizes f:

$$\text{Max} \quad \left[\left(2 + \frac{2}{\sqrt{5}}\,d - 1\right)^2 + \left(\frac{1}{2} + \frac{1}{\sqrt{5}}\,d\right)\right]$$

$$0 < d \leqslant d_{\text{Max}}$$

It is obvious that the term in square brackets increases without limit as d increases. Hence $d^* = d_{Max}$. d_{Max} is obtained in the following way. Since it is apparent that the gradient direction will intersect constraint (ii), the expressions for x_1 and x_2 may be inserted into this constraint

$$6\left(\frac{2}{\sqrt{5}}\,d + 2\right) + \left(\frac{1}{2} + \frac{1}{\sqrt{5}}\,d\right) \leqslant 16.$$

Hence, $$d \leqslant \frac{7}{26}\,\sqrt{5} = 0.6 = d_{\text{Max}}.$$

This procedure is unnecessary, since it has already been determined that constraint (ii) limits further search in the gradient direction. However, knowing d_{Max}, we can identify the point where the gradient direction intersects constraint (ii) as $\bar{x}^1 = (33/13, 10/13)$.

At this point we evaluate

$$\left(\frac{\partial f}{\partial x_1}\right)_1 = 2(x_1^1 - 1) = 2(\frac{20}{13}) = \frac{40}{13}$$

$$\left(\frac{\partial f}{\partial x_2}\right)_1 = 1.$$

Again the partial derivatives are not zero; the gradient direction is (0.93, 0.37). As shown on part (b) of Figure 5.6A, this direction would lead the search outside the feasible region. Hence, we ask if its projection onto the feasible region makes a nonobtuse angle. It does. This projection is shown on Figure 5.6A. The constrained ascent is along constraint (ii), proceeding in the direction of increasing x_2. We again apply equation (5.8)

$$\text{Max} \quad \left[\left(\frac{33}{13} + 0.93d - 1\right)^2 + \left(\frac{10}{13} + 0.37d\right)\right].$$
$$0 < d \leq d_{\text{Max}}$$

Although we could proceed as before to evaluate d_{Max}, it is obvious that this will occur at the intersection of constraints (i) and (ii). Hence $\bar{x}^2 = (7/3, 2)$ and

$$\left(\frac{\partial f}{\partial x_1}\right)_2 = 2(x_1^2 - 1) = 8/3$$

$$\left(\frac{\partial f}{\partial x_2}\right)_2 = 1.$$

The gradient direction is $(0.94, 0.35)$ as shown on Figure 5.6A. This direction points outside of the feasible region as before. Only in this case, there is no projection onto the feasible region which makes a nonobtuse angle with the gradient direction. Hence the point $\bar{x}^2 = (7/3, 2)$ is a local maximum. The value of f is 3 7/9.

Suppose now we had initiated our search at the point $(1, 1)$. Here

$$\left(\frac{\partial f}{\partial x_1}\right)_0 = 2(1 - 1) = 0$$

$$\left(\frac{\partial f}{\partial x_2}\right)_0 = 1.$$

The gradient direction is the vertical line from the point $(1, 1)$. This is shown on part (a) of Figure 5.6B. This direction will lead to the next test point, $\bar{x}^1 = (1, 2)$.

The partial derivatives are

$$\left(\frac{\partial f}{\partial x_1}\right)_1 = 2(1 - 1) = 0$$

$$\left(\frac{\partial f}{\partial x_2}\right)_1 = 1.$$

Again the gradient direction is the vertical line from the point $(1, 2)$. Now this line points outside the feasible region, and we must resort to a projection onto the feasible region. Since the gradient is normal to the feasible region, however, no direction is provided and the gradient technique would identify the point $\bar{x}^1 = (1, 2)$ as a relative maximum. Actually it is a constrained saddle point since the function could be increased if x_1 were moved in either direction. Nonetheless, in one step beginning at $\bar{x}^0 = (1, 1)$, the gradient technique identifies (incorrectly) the point $(1, 2)$ as a relative maximum; $f = 2$.

In a third attempt to use the gradient method to solve the problem, select $\bar{x}^0 = (0, 0)$. Here

$$\left(\frac{\partial f}{\partial x_1}\right)_0 = 2(x_1 - 1) = -2$$

$$\left(\frac{\partial f}{\partial x_2}\right)_0 = 1.$$

The gradient direction is $\left(\dfrac{-2}{\sqrt{5}}, \dfrac{1}{\sqrt{5}}\right)$, pointing out of the feasible region. Its projection onto the feasible region does make a nonobtuse angle with the x_2 axis. Hence, by proceeding along the x_2-axis in a positive direction, we are following the constrained ascent direction.

Equation (5.8) is again in order. However, it is obvious once again that the function increases without limit as the distance from the origin along the x_2-axis increases. The maximum increase is to the point $\bar{x}^1 = (0, 2)$ beyond which constraint (i) would be violated. At $\bar{x}^1$,

$$\left(\frac{\partial f}{\partial x_1}\right)_1 = 2(0 - 1) = -2$$

$$\left(\frac{\partial f}{\partial x_2} \right)_1 = 1.$$

The gradient direction is still $\left(\dfrac{-2}{\sqrt{5}} , \dfrac{1}{\sqrt{5}} \right)$, except that at $\bar{x}^1$ its projection on the feasible region makes an obtuse angle with the gradient direction. Hence this point is identified as a relative maximum; $f = 3$.

This example has served to illustrate the gradient technique applied to a problem with linear inequality constraints. It has further demonstrated that the use of necessary conditions to identify an extremum turns up local maxima and saddle points in addition to the global maximum. It is difficult to imagine a simpler constrained optimization problem than that posed in example 5.1. Yet three different starting points gave three different answers. In more complicated problems, therefore, it is always necessary to apply the steepest ascent technique several times, with different starting points. If different extrema are located, the desired answer is the one with the largest value for the objective function. Thus, in the example, $\bar{x}^* = (7/3, 2)$ is the global maximizing point and $f = 3\ 7/9$ is the global maximum of the function.

It should again be noted that the gradient projection method provides a simple procedure for selecting a direction at a boundary, which insures that the objective function increases. It is possible to determine the direction of *steepest* ascent at a boundary if one is willing to solve a related optimization problem at each encounter with the boundary. In order to illustrate the nature of this approach, consider a point on the boundary $\bar{x}^k$ where the $\ell\underline{th}$ constraint is active:

$$\sum_{j=1}^n a_{\ell j} x_j^k = b_\ell. \tag{5.9}$$

The others are assumed inactive for the sake of illustration.

If $\{r_j^k\}$ denotes the components of the direction of search from $\bar{x}^k$, then along this direction

$$x_j = x_j^k + d \ r_j^k \qquad j = 1, 2, \ldots, n.$$

If these values are substituted into the $\ell^{\underline{th}}$ constraint equation, we have

$$\sum_{j=1}^{n} a_{\ell j}(x_j^k + d \ r_j^k) \leqslant b_\ell. \tag{5.10}$$

Subtraction of (5.9) from (5.10) shows

$$\sum_{j=1}^{n} a_{\ell j} \ r_j^k \leqslant 0. \tag{5.11}$$

If more than one constraint is active, it is necessary to impose inequalities such as (5.11) for each operative constraint.

We now consider the same problem for determining $\bar{r}^k$ as we did in the unconstrained case, namely

$$\text{Max} \sum_{j=1}^{n} \left(\frac{\partial f}{\partial x_j}\right)_k r_j^k \tag{5.12}$$

$$\text{Subject to} \sum_{j=1}^{n} (r_j^k)^2 = 1. \tag{5.13}$$

Equations (5.11), (5.12) and (5.13) constitute in themselves a nonlinear programming problem whose solution determines the steepest constrained ascent direction. Notice that the nonlinearity stems from the normalization criterion on the $\{r_j\}$, equation (5.13). One suggestion is to replace this requirement by the less restrictive but linear conditions

$$- 1 \leqslant r_j^k \leqslant 1 \qquad j = 1, 2, \ldots, n. \quad (5.14)$$

Then the problem is a linear programming one that can be solved by the simplex algorithm.

A second way of dealing with the problem is to solve the following quadratic programming problem:

$$\text{Min } z = \sum_{j=1}^{n} (u_j)^2$$

$$\text{Subject to} \quad \sum_{j=1}^{n} \left(\frac{\partial f}{\partial x_j} \right)_k u_j = 1$$

$$\sum_{j=1}^{n} a_{\ell j} u_j \leqslant 0.$$

Hadley (*11*) has shown that the solution to this problem is equivalent to the problem posed by (5.11), (5.12) and (5.13) with

$$u_j^* = \frac{\left(r_j^k \right)^*}{\sum_{i=1}^{n} \left(\frac{\partial f}{\partial x_i} \right)_k (r_i^k)^*}$$

While either of these alternatives may provide a search direction at a boundary similar to that obtained by gradient projection, they require more computations to identify this direction. Nonetheless, they have been used to good advantage in solving many engineering problems.

5.2.3 Nonlinear Inequality Constraints When the constraints are nonlinear, the problem of selecting a feasible steepest ascent path is more difficult. In this case, the parameter optimization problem is stated as

$$\text{Max } f(x_1, x_2, \ldots, x_n) \tag{5.15}$$

$$g_i(x_1, x_2, \ldots, x_n) \leqslant 0 \qquad i = 1, 2, \ldots, m \tag{5.16}$$

$$x_j \geqslant 0 \qquad j = 1, 2, \ldots, n. \tag{5.17}$$

and we will again assume that the region of feasible solutions identified by the constraints is convex. Also, we assume f and all the $\{g_i\}$ have continuous first derivatives.

As in the previous steepest ascent techniques, this problem is solved by following a directed path through the region of feasible solutions. At interior test points, the procedure for generating a search direction is similar to that used in the two previous cases. It is at boundaries of the feasible region where the difficulties are encountered.

In order to illustrate the complications which arise at a nonlinear boundary, consider the situation depicted in Figure 5.7. At the test point, the constraint is active and

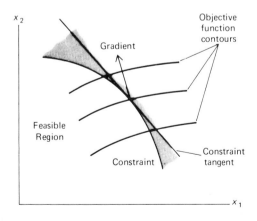

Figure 5.7

the gradient points out of the feasible region. Thus we must locate the constrained direction of steepest ascent. If we proceed as in the last section, we seek the direction $\bar{r}^k$ which maximizes the directional derivative of f. If the ℓth constraint is active, $(\bar{r}^k)^*$ is the solution to the problem:

$$\text{Max} \sum_{j=1}^{n} \left(\frac{\partial f}{\partial x_j} \right)_k r_j^k$$

$$\sum_{j=1}^{n} (r_j^k)^2 = 1$$

$$g_\ell(x_1^k + d\ r_1^k, \ldots, x_n^k + d\ r_n^k) \leqslant 0.$$

Since there is no simple solution to this problem in general, the approach taken is to simplify the problem to the extent that a solution can be obtained but not so drastically that the computed direction fails to yield function improvement.

The most common approximation is to replace the constraint equation by its tangent at $\bar{x}^k$. This replaces (5.16) by

$$\sum_{j=1}^{n} \left(\frac{\partial g_i}{\partial x_j} \right)_k (x_j - x_j^k) \leqslant 0$$

or, in terms of r_j^k

$$\sum_{j=1}^{n} \left(\frac{\partial g_i}{\partial x_j} \right)_k r_j^k \leqslant 0. \tag{5.18}$$

This equation is then identical to (5.11) and either of the solution procedures discussed at the end of section 5.2.2 could be used to determine a search direction.

The difficulty with this method is that the search direction obtained by replacing the constraint equation with its tangent may not be feasible. This is illustrated in Figure 5.8. The steepest constrained path computed by the linearization procedure is along the constraint tangent, which leads outside of the feasible region. If the $(k + 1)^{st}$ test point is to be feasible, the above procedure must be modified.

The most common modification is to require that the constraint tangent not merely be less than zero but be less than some value $- \beta_i$, i.e.,

$$\sum_{j=1}^{n} \left(\frac{\partial g_i}{\partial x_j} \right)_k r_j^k \leqslant - \beta_i.$$

As shown in Figure 5.8, this forces the search direction into the region of feasible solutions, provided β_i is large enough. Care must be exercised in the selection of β_i. Too small a value would cause the search direction to point out of the feasible region. Too large a value may cause the search direction to point in a direction of function decrease. The recommended procedure of selecting β_i is to begin with a small value that leads out of the feasible region and iteratively increase β_i until it does lead to an improved test point inside the feasible region.

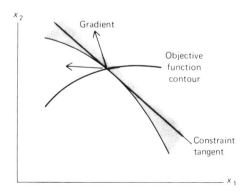

Figure 5.8

Example 5.2 Find the rectangle of largest area which fits inside of the unit circle.

Solution We wish to maximize the area

$$z = 4x_1x_2$$

subject to

$$x_1^2 + x_2^2 \leqslant 1.$$

Let us begin the search at $(1/2, 0)$.

Here

$$\left(\frac{\partial f}{\partial x_1}\right) = 4x_2 = 0 \qquad \left(\frac{\partial f}{\partial x_2}\right) = 4x_1 = 2.$$

The gradient direction points vertically. This direction violates no constraint so that we increase x_2. It is obvious that we continue to increase x_2 until the constraint is reached at the point $\left(\frac{1}{2}, \frac{\sqrt{3}}{2}\right)$. Here

$$\left(\frac{\partial f}{\partial x_1}\right) = 4x_2 = 2\sqrt{3}; \left(\frac{\partial f}{\partial x_2}\right) = 4x_1 = 2.$$

Hence the gradient direction points out of the feasible region, and we must settle for a constrained ascent.

The constraint derivatives at the test point are

$$\left(\frac{\partial g}{\partial x_1}\right) = 2x_1 = 1 \qquad \left(\frac{\partial g}{\partial x_2}\right) = 2x_2 = \sqrt{3}$$

Hence the problem to find a constrained ascent direction is

$$\text{Max } 2\sqrt{3}\, r_1 + 2\, r_2$$

subject to

$$-1 \leqslant r_1 \leqslant 1$$

$$-1 \leqslant r_2 \leqslant 1$$

$$r_1 + \sqrt{3}\ r_2 \leqslant \beta.$$

Although this problem could be solved by the simplex method it is informative to plot the feasible region and objective function contours. This is done on Figure 5.9. It is obvious that the solution points occur at the vertex formed at the right-hand-most corner of the feasible region. These vertices are labeled by dark circles on Figure 5.9.

For $\beta = 0$, the best directions are (un-normalized)

$$r_1 = 1\ ,\ r_2 = -\ 1/\sqrt{3}$$

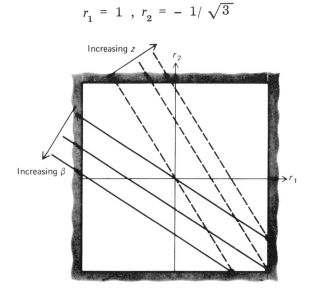

Figure 5.9

Thus, along this path

$$x_1 = 1/2 + d\ ,\ x_2 = \frac{\sqrt{3}}{2} - \frac{d}{\sqrt{3}}.$$

If these values are substituted into the constraint equation, we find that

$$(1/2 + d)^2 + \left(\frac{\sqrt{3}}{2} - \frac{d}{\sqrt{3}} \right)^2 \leqslant 1.$$

which implies $\dfrac{4}{3} d^2 \leqslant 0.$

This can only be true for $d = 0$. Hence, the move along the constraint tangent leads out of the feasible region. We must try again with larger β.

For $\beta = -1/2$, the solution to the linear programming problem is

$$r_1 = 1 \ , \ r_2 = -\sqrt{3}/2.$$

Thus, along the search path,

$$x_1 = 1/2 + d \ , \ x_2 = \left(\frac{\sqrt{3}}{2} - \frac{\sqrt{3}}{2} d \right).$$

The constraint equation

$$(1/2 + d)^2 + \left(\frac{\sqrt{3}}{2} - \frac{\sqrt{3}}{2} d \right)^2 \leqslant 1$$

now implies $-\dfrac{1}{2} d + \dfrac{7}{4} d^2 \leqslant 0.$

This permits d to have a value as high as $2/7$. Thus, this move is feasible.

The objective function becomes

$$z = 4(1/2 + d) \left(\frac{\sqrt{3}}{2} - \frac{\sqrt{3}}{2} d \right)$$
$$= 2\sqrt{3} \left(\frac{1}{2} + \frac{1}{2} d - d^2 \right).$$

This function is maximized for $d = 1/4$ within the feasible region. Hence, the next trial point is $(3/4, 3\sqrt{3}/8)$.

Here, the function derivatives are

$$\frac{\partial f}{\partial x_1} = 4x_2 = \frac{3}{2}\sqrt{3}; \quad \left(\frac{\partial f}{\partial x_2}\right) = 4x_1 = 3.$$

The gradient direction defined by these derivatives is feasible, and the search continues as before, ultimately locating the extreme point $(1/\sqrt{2}, 1/\sqrt{2})$.

It should be pointed out that the second choice of $\beta = 1/2$ was completely arbitrary. Generally a smaller increase is advised.

The problem described in this section is the most general of the parameter optimization problems we have considered. There is no single optimizing routine which is most efficient for every problem in this class. However, much work is being carried out to develop more efficient methods, and there is every reason to believe that newer and more powerful methods for dealing with the general optimization problem will be developed.

5.3 SECOND-ORDER TECHNIQUES

One shortcoming of gradient methods based on the steepest ascent principle is their inability to ascend a sharp ridge rapidly. Since it is not always possible to deal with near spherical response surfaces, a number of optimization techniques have been proposed which try to circumvent movement entirely in the gradient direction. To illustrate this, let us represent an *arbitrary* two-dimensional objective function in terms of a quadratic function in which the c's and q's are not given explicitly:

$$f = c_0 + c_1 x_1 + c_2 x_2 - \frac{1}{2}(q_{11}x_1^2 + 2q_{12}x_1 x_2 + q_{22}x_2^2). \tag{5.19}$$

The reason for the double subscripts on the quadratic

terms is that the two-dimensional objective function is a special case of an n-dimensional quadratic function which can be represented as

$$f = c_0 + \sum_{i=1}^{n} c_i x_i - \frac{1}{2} \sum_{j=1}^{n} \sum_{i=1}^{n} q_{ij} x_i x_j. \quad (5.20)$$

In order to insure that the function f is concave, we will require that the quadratic term be positive for all non-zero values of x_i, x_j.

The contour plot of f is depicted on Figure 5.10 along with the gradient direction at $\bar{x}^0$. Note that the gradient direction does not point to the extremum. To show this, consider now that at $\bar{x}^0$ the partial derivatives are evaluated. Symbolically they are given as

$$\left(\frac{\partial f}{\partial x_1}\right)_0 = c_1 - q_{11} x_1^0 - q_{12} x_2^0 \quad (5.21)$$

$$\left(\frac{\partial f}{\partial x_2}\right)_0 = c_2 - q_{12} x_1^0 - q_{22} x_2^0. \quad (5.22)$$

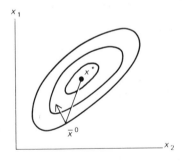

Figure 5.10

For the objective function given in (5.19), we can identify $\bar{x}^*$ symbolically by the techniques of chapter 1

$$\left(\frac{\partial f}{\partial x_1} \right)_* = c_1 - q_{11}x_1^* - q_{12}x_2^* = 0 \qquad (5.23)$$

$$\left(\frac{\partial f}{\partial x_2} \right)_* = c_2 - q_{12}x_1^* - q_{22}x_2^* = 0. \qquad (5.24)$$

Subtraction of (5.23) from (5.21) and (5.24) from (5.22) yields

$$\left(\frac{\partial f}{\partial x_1} \right)_0 = q_{11}(x_1^* - x_1^0) + q_{12}(x_2^* - x_2^0) \qquad (5.25)$$

$$\left(\frac{\partial f}{\partial x_2} \right)_0 = q_{12}(x_1^* - x_1^0) + q_{22}(x_2^* - x_2^0). \qquad (5.26)$$

Equations (5.25) and (5.26) form a linear system of equations which may be solved by Cramer's Rule, provided $q_{11}q_{22} - q_{12}^2 \neq 0$:

$$(x_1^* - x_1^0) = \frac{q_{22}\left(\frac{\partial f}{\partial x_1} \right)_0 - q_{12}\left(\frac{\partial f}{\partial x_2} \right)_0}{q_{11}q_{22} - q_{12}^2} \qquad (5.27)$$

$$(x_2^* - x_2^0) = \frac{q_{11}\left(\frac{\partial f}{\partial x_2} \right)_0 - q_{12}\left(\frac{\partial f}{\partial x_1} \right)_0}{q_{11}q_{22} - q_{12}^2} \qquad (5.28)$$

It is obvious from these last two equations that the gradient direction is not pointed toward $\bar{x}^*$ unless $q_{12} = 0$— i.e., unless there are no interactions. Since in general interactions can be expected, it would be advantageous to move in the direction suggested by equations (5.27) and (5.28).

In most real situations, the coefficients $\{q_{ij}\}$ are not known beforehand. If they were known or could be evaluated, then equations (5.27) and (5.28) could be used, not merely to suggest a direction but to suggest the next test point itself. Most second-order methods provide ways of obtaining the $\{q_{ij}\}$ as the actual search routine proceeds.

To provide an appreciation for what is involved in the determination of the $\{q_{ij}\}$ during the solution process itself, consider the maximization of the general two-dimensional quadratic by a second-order technique. At the initial point $\bar{x}^0$, the derivatives are evaluated and (5.21) and (5.24) hold. Since we lack information on the $\{q_{ij}\}$, at this point we choose to follow the gradient direction and select the distance d from $\bar{x}^0$ which maximizes f in this direction. That is,

$$\underset{\{d\}}{\text{Max}} \; f \left[x_1^0 + \left(\frac{\partial f}{\partial x_1} \right)_0 d \; , \; x_2^0 + \left(\frac{\partial f}{\partial x_2} \right)_0 d \right]. \tag{5.29}$$

In the unconstrained case, this is found by setting the derivative of f with respect to d equal to zero. This derivative is obtained by application of chain rule differentiation of f

$$\left(\frac{\partial f}{\partial x_1} \right)_1 \left(\frac{\partial f}{\partial x_1} \right)_0 + \left(\frac{\partial f}{\partial x_2} \right)_1 \left(\frac{\partial f}{\partial x_2} \right)_0 = 0. \tag{5.30}$$

The partial derivatives appearing first in the two terms in (5.30) are evaluated at $\bar{x}^1$, the trial point identified by the selection of d. Also at $\bar{x}^1$ we have symbolically

$$\left(\frac{\partial f}{\partial x_1}\right)_1 = c_1 - q_{11}x_1^1 - q_{12}x_2^1 \qquad (5.31)$$

$$\left(\frac{\partial f}{\partial x_2}\right)_1 = c_2 - q_{12}x_1^1 - q_{22}x_2^2. \qquad (5.32)$$

Equations (5.21), (5.22), (5.30), (5.31) and (5.32) constitute a set of five equations in the five unknowns c_1, c_2, q_{11}, q_{12}, and q_{22}. Although it is possible to solve for all five unknowns, only the $\{q_{ij}\}$ are needed to locate $\bar{x}^*$. Hence c_1 and c_2 may be eliminated from the equations. To see how this is done, subtract (5.31) from (5.21) and (5.32) from (5.22):

$$\left(\frac{\partial f}{\partial x_1}\right)_0 - \left(\frac{\partial f}{\partial x_1}\right)_1 = q_{11}(x_1^1 - x_1^0) + q_{12}(x_2^1 - x_2^0)$$
$$(5.33)$$

$$\left(\frac{\partial f}{\partial x_2}\right)_0 - \left(\frac{\partial f}{\partial x_2}\right)_1 = q_{12}(x_1^1 - x_1^0) + q_{22}(x_2^1 - x_2^0)$$
$$(5.34)$$

To simplify notation, let

$$\triangle x_1^k = x_1^k - x_1^{k-1}$$
$$k = 1, 2, \ldots \quad (5.35)$$
$$\triangle x_2^k = x_2^k - x_2^{k-1}$$

and

$$\triangle G_1^k = \left(\frac{\partial f}{\partial x_1}\right)_k - \left(\frac{\partial f}{\partial x_1}\right)_{k-1}$$
$$k = 1, 2, \ldots$$
$$\triangle G_2^k = \left(\frac{\partial f}{\partial x_2}\right)_k - \left(\frac{\partial f}{\partial x_2}\right)_{k-1}$$

So that equations (5.33) and (5.34) become

$$\triangle x_1^1 \ q_{11} + \triangle x_2^1 \ q_{12} = - \triangle G_1^1 \qquad (5.36)$$

$$\triangle x_1^1 \ q_{12} + \triangle x_2^1 \ q_{22} = - \triangle G_2^1. \qquad (5.37)$$

Remember that the $\triangle x$'s and $\triangle G$'s are quantities which can be computed from the test points and the partial derivative values at the test points. Only q_{11}, q_{12}, and q_{22} are unknown in (5.36) and (5.37).

Substitution of (5.33) and (5.34) into (5.30) leads to the following:

$$\left(\frac{\partial f}{\partial x_1} \right)_1 \left[\left(\frac{\partial f}{\partial x_1} \right)_1 + q_{11} \triangle x_1^1 + q_{12} \triangle x_2^1 \right]$$

$$+ \left(\frac{\partial f}{\partial x_2} \right)_1 \left[\left(\frac{\partial f}{\partial x_2} \right)_1 + q_{12} \triangle x_1^1 + q_{22} \triangle x_2^1 \right] = 0$$

or

$$\left(\frac{\partial f}{\partial x_1} \right)_1 \triangle x_1^1 \ q_{11} + \left[\left(\frac{\partial f}{\partial x_1} \right)_1 \triangle x_2^1 + \left(\frac{\partial f}{\partial x_2} \right)_1 \triangle x_1^1 \right] q_{12}$$

$$+ \left(\frac{\partial f}{\partial x_2} \right)_1 \triangle x_2^1 \ q_{22} = - \left[\left(\frac{\partial f}{\partial x_1} \right)_1^2 + \left(\frac{\partial f}{\partial x_2} \right)_1^2 \right]. \qquad (5.38)$$

Equations (5.36), (5.37) and (5.38) constitute a linear set of equations in the $\{q_{ij}\}$ and solution is possible by Cramer's Rule. With the $\{q_{ij}\}$ determined, (5.27) and (5.28) can be used to locate $\bar{x}^*$.

To evaluate the $\{q_{ij}\}$, we required knowledge of two

test points, the values of the partial derivatives at these points, and the condition that the second test point be located by maximizing the objective function in the gradient direction emanating from the first test point. If the objective function were n-dimensional, n such pieces of information would be required.

Example 5.3

$$\text{Maximize } x_1 + x_2 - \frac{1}{2}(x_1^2 + 2x_1x_2 + 2x_2^2)$$

Solution The answer is $x_1^* = 1$, $x_2^* = 0$.

Begin the second-order technique at $(1, 1)$

$$\left(\frac{\partial f}{\partial x_1}\right)_0 = 1 - x_1 - x_2 = 1 - 1 - 1 = -1$$

$$\left(\frac{\partial f}{\partial x_2}\right)_0 = 1 - x_1 - 2x_2 = 1 - 1 - 2 = -2.$$

Let the first search be in gradient direction

$$\underset{\{d\}}{\text{Max}} \left[(1 - d) + (1 - 2d) - \frac{1}{2}\left((1 - d)^2 \right.\right.$$
$$\left.\left. + 2(1 - d)(1 - 2d) + 2(1 - 2d)^2 \right) \right].$$

By standard means, we find

$$d^* = 5/13 \ , \ x_1^1 = 8/13 \ , \ x_2^1 = 3/13$$

so that $\Delta x_1^1 = -5/13 \quad \Delta x_2^1 = -10/13$

and

$$\left(\frac{\partial f}{\partial x_1}\right)_1 = 1 - x_1 - x_2 = 1 - 8/13 - 3/13 = 2/13$$

$$\left(\frac{\partial f}{\partial x_2}\right)_1 = 1 - x_1 - 2x_2 = 1 - 8/13 - 6/13 = -1/13.$$

Hence $\triangle G_1^1 = 15/13$ and $\triangle G_2^1 = 25/13.$

Equations (5.36), (5.37) and (5.38) are now invoked

$$-\frac{15}{13} q_{11} - \frac{10}{13} q_{12} \qquad\qquad = -\frac{15}{13}$$

$$-\frac{5}{13} q_{12} - \frac{10}{13} q_{22} = -\frac{25}{13}$$

$$-\frac{10}{(13)^2} q_{11} - \frac{15}{(13)^2} q_{12} + \frac{10}{(13)^2} q_{22} = -\frac{5}{(13)^2}.$$

These yield $q_{11} = 1$, $q_{12} = 1$, $q_{22} = 2$, as expected.
Application of (5.27) and (5.28) correctly obtains $\bar{x}^*$:

$$x_1^* - 1 = \frac{2(-1) - (1)(-2)}{1} \quad ; \quad x_1^* = 1$$

$$x_2^* - 1 = \frac{1(-2) - 1(-1)}{1} \quad ; \quad x_2^* = 0.$$

Although it is possible to utilize the approach described above in more general situations, it is not difficult to see that the gradients at each test point would have to be stored and that the final computations to obtain the $\{q_{ij}\}$ could be quite involved. Moreover, unless some care is taken in the selection of the search directions, it is possible that these computations could be subject to considerable numerical error. Consequently, a more formalized search plan employs the concept of *conjugate directions*.

5.3.1 Conjugate Directions The idea behind the conjugate direction method is the generation of search directions which are mutually orthogonal (perpendicular). This insures that the equations for the $\{q_{ij}\}$ will have a unique solution. Moreover, it makes it possible to utilize the gradients at the test points as they are generated in a sequential solution for the $\{q_{ij}\}$.

In order to illustrate the technique, consider maximization of a function f approximated as a quadratic

$$\text{Max} \quad z \;=\; f(\bar{x}) \;=\; \sum_{i=1}^{n} c_i x_i \;-\; \frac{1}{2} \sum_{j=1}^{n} \sum_{i=1}^{n} q_{ij} x_i x_j. \tag{5.39}$$

For the moment, consider that the coefficients in this function are known. Later we will show how to proceed when they must be evaluated during the search.

Let the search direction at the kth test point be identified as

$$\bar{r}^k \;=\; (r_1^k, \; r_2^k, \; \ldots \; r_n^k). \tag{5.40}$$

Then the components of the $(k + 1)^{st}$ test point are

$$x_i^{k+1} \;=\; x_i^k \;+\; d_k^* \; r_i^k \tag{5.41}$$

where d_k^* is the optimum distance in the direction $\bar{r}^k$ from x^k. That is, d_k^* maximizes

$$z \;=\; f(\bar{x}^k \;+\; d_k \; \bar{r}^k).$$

Thus d_k^* must satisfy

$$\frac{\partial z}{\partial d_k} \;=\; \sum_{\ell=1}^{n} \left(\frac{\partial f}{\partial x_\ell} \right)_{k+1} r_\ell^k \;=\; 0. \tag{5.42}$$

Equation (5.42) is equivalent to saying that the function gradient at $\bar{x}^{k+1}$ is perpendicular to the search direction from $\bar{x}^k$. If $\bar{r}^{k+1}$ is the search direction from $\bar{x}^{k+1}$, the $(k+2)^{nd}$ test point, $\bar{x}^{k+2}$, has

$$x_i^{k+2} = x_i^{k+1} + d_{k+1}^* \, r_i^{k+1} \qquad i = 1, \, 2, \, \ldots, \, n. \tag{5.43}$$

Substitution of (5.41) into this expression yields

$$x_i^{k+2} = x_i^k + d_k^* \, r_i^k + d_{k+1}^* \, r_i^{k+1} \quad i = 1, \, 2, \, \ldots, \, n.$$

Continuing in this way, we obtain for the nth test point

$$x_i^n = x_i^k + \sum_{j=k}^{n} d_j^* \, r_i^j \qquad i = 1, \, 2, \, \ldots, \, n. \tag{5.44}$$

Since the partial derivative of the objective function with respect to x_ϱ at the nth test point is

$$\left(\frac{\partial f}{\partial x_\varrho} \right)_n = c_\varrho - \sum_{i=1}^{n} q_{\varrho_i} \, x_i^n \, ,$$

(5.44) can be used to obtain

$$\left(\frac{\partial f}{\partial x_\varrho} \right)_n = c_\varrho - \sum_{i=1}^{n} q_{\varrho_i} \left[x_i^k + \sum_{j=k}^{n} d_j^* \, r_i^j \right] . \tag{5.45}$$

The same partial derivative at the kth test point is

$$\left(\frac{\partial f}{\partial x_\varrho} \right)_k = c_\varrho - \sum_{i=1}^{n} q_{\varrho_i} \, x_i^k . \tag{5.46}$$

Subtraction of (5.46) from (5.45) gives

$$\left(\frac{\partial f}{\partial x_\ell} \right)_n = \left(\frac{\partial f}{\partial x_\ell} \right)_k - \sum_{i=1}^{n} q_{\ell_i} \sum_{j=k}^{n} d_j^* \, r_i^j. \quad (5.47)$$

If we multiply (5.47) by r_ℓ^{k-1} and sum the resultant equation from $\ell = 1$ to n, we find

$$\sum_{\ell=1}^{n} r_\ell^{k-1} \left(\frac{\partial f}{\partial x_\ell} \right)_n = \sum_{\ell=1}^{n} r_\ell^{k-1} \left(\frac{\partial f}{\partial x_\ell} \right)_k$$

$$- \sum_{\ell=1}^{n} r_\ell^{k-1} \sum_{i=1}^{n} q_{\ell_i} \sum_{j=k}^{n} d_j^* \, r_i^j. \quad (5.48)$$

The first summation on the right-hand side of (5.48) vanishes by virtue of (5.42). If the directions are selected so that

$$\text{for } j \neq k - 1, \quad \sum_{\ell=1}^{n} \sum_{i=1}^{n} q_{\ell_i} \, r_i^j \, r_\ell^{k-1} = 0. \quad (5.49)$$

Then the left-hand side of (5.48) equals zero

$$\sum_{\ell=1}^{n} r_\ell^{k-1} \left(\frac{\partial f}{\partial x_\ell} \right)_n = 0.$$

Because the $(k-1)^{st}$ directions are not zero in general for all k, it must be that

$$\left(\frac{\partial f}{\partial x_\ell} \right)_n = 0 \qquad \ell = 1, 2, \ldots, n, \quad (5.50)$$

Hence, at the nth test point, the partial derivatives of the objective function with respect to all the independent variables vanish. Since the objective function is concave, this condition is both necessary and sufficient for the optimum. Hence $\bar{x}^n$ is the desired solution to the problem.

The condition which insures that $\bar{x}^n$ is the maximum point is (5.49). The directions computed from this condition are the conjugate directions. It should be noted that there is not a unique set of conjugate directions. Thus, *any set of* directions which satisfy condition (5.49) can be used in order to insure that $\bar{x}^n$ maximizes the objective function.

Example 5.4 Resolve the problem of example 5.3, which was to maximize

$$\text{Max } z = f = x_1 + x_2 - \frac{1}{2} (x_1^2 + 2x_1 x_2 + 2x_2^2).$$

Solution We first compute the conjugate directions, assuming the quadratic coefficients are known. By (5.49) for $n = 2$

$$\sum_{\ell=1}^{2} \sum_{i=1}^{2} q_{\ell_i} r_i^0 r_\ell^1 = r_1^0 r_1^1 + r_1^0 r_2^1 + r_2^0 r_1^1 + 2r_2^0 r_2^1 = 0$$

It is apparent that there are an infinite number of combinations of the various directions which satisfy this equation. Some normalized sets are

$(r_1^0 = 0, \ r_2^0 = 1)$, $(r_1^1 = 2/\sqrt{5}, \ r_2^1 = -1/\sqrt{5})$

$(r_1^0 = 1, \ r_2^0 = 0)$, $(r_1^1 = 1/\sqrt{2}, \ r_2^1 = -1/\sqrt{2})$

$(r_1^0 = 1/\sqrt{2}, \ r_2^0 = 1/\sqrt{2})$, $(r_1^1 = \dfrac{3}{\sqrt{13}}, \ r_2^1 = \dfrac{-2}{\sqrt{13}})$.

We choose the third of these sets and begin the search at the point used in the previous example, (1,1)., The conjugate direction provides

$$x_1 = 1 + d_1 (1/\sqrt{2})$$

$$x_2 = 1 + d_1 (1/\sqrt{2}).$$

The objective function is

$$f(\bar{x}^0 + d_1 \bar{r}^0) = 2 \left(1 + \frac{d_1}{\sqrt{2}}\right) - \frac{5}{2} \left(1 + \frac{d_1}{\sqrt{2}}\right)^2$$

and

$$d_1^* = -\frac{3}{5} \sqrt{2}.$$

Hence

$$x_1^1 = 2/5, \; x_2^1 = 2/5.$$

The second set of conjugate directions provides the following values for x_1 and x_2 along the search path

$$x_1 = 2/5 + d_2 \left(\frac{3}{\sqrt{13}}\right)$$

$$x_2 = 2/5 + d_2 \left(\frac{-2}{\sqrt{13}}\right)$$

and the objective function is

$$f(\bar{x}^1 + d_2 \bar{r}^1) = \frac{2}{5} + \frac{1}{\sqrt{13}} d_2 - \frac{5}{26} d_2^2.$$

Hence

$$d_2^* = \frac{1}{5} \sqrt{13}$$

and

$$x_1^2 = 1, \; x_2^2 = 0$$

which is the optimum solution.

This example is somewhat artificial in the sense that the quadratic coefficients were assumed known. In an actual optimization situation, these coefficients will not be known (or else one could use other techniques to solve the problem). Thus, it will be necessary to employ procedures which identify the conjugate directions in the absence of this information. However, as we saw earlier, it is possible to use the values of the gradients at the test points to identify the $\{q_{ij}\}$. It is similarly possible to use the gradient values to identify the conjugate gradients.

The sequential development of the conjugate gradients proceeds as follows. An initial point $\bar{x}^0$ is selected and the gradient direction computed. The initial search is in this direction, so that

$$r_i^0 = \left(\frac{\partial f}{\partial x_i} \right)_0 \qquad\qquad i = 1, 2, \ldots, n.$$

The search is terminated in this direction at $\bar{x}^1$, by selection of the distance from $\bar{x}^0$ in the direction $\bar{r}^0$ which maximizes the objective function. At x^1 the partial derivatives of f with respect to the $\{x_i\}$ are computed.

By the arguments made in the preceding section, the partial derivative of f with respect to x_ϱ at $\bar{x}^1$ can be written as

$$\left(\frac{\partial f}{\partial x_\varrho} \right)_1 = c_\varrho - \sum_{i=1}^{n} q_{\varrho_i} x_i^0 - d_0^* \sum_{i=1}^{n} q_{\varrho_i} r_i^0$$

whereas the same derivative at $\bar{x}^0$ is symbolically

$$\left(\frac{\partial f}{\partial x_\varrho} \right)_0 = c_\varrho - \sum_{i=1}^{n} q_{\varrho_i} x_i^0.$$

Hence

$$\left(\frac{\partial f}{\partial x_\ell}\right)_1 - \left(\frac{\partial f}{\partial x_\ell}\right)_0 = -d_0^* \sum_{i=1}^{n} q_{\ell_i} r_i^0. \qquad (5.51)$$

Multiplication of (5.51) by $\bar{r}_\ell^1$ and summation from $\ell = 1$ to n gives

$$\sum_{\ell=1}^{n} r_\ell^1 \left(\frac{\partial f}{\partial x_\ell}\right)_1 - \sum_{\ell=1}^{n} r_\ell^1 \left(\frac{\partial f}{\partial x_\ell}\right)_0 = -d_0^* \sum_{\ell=1}^{n} r_\ell^1 \sum_{i=1}^{n} q_{\ell_i} r_i^0.$$

If the directions $\bar{r}^0$ and $\bar{r}^1$ are to be conjugate, the right-hand side of this equation must vanish, so that

$$\sum_{\ell=1}^{n} r_\ell^1 \left(\frac{\partial f}{\partial x_\ell}\right)_1 - \sum_{\ell=1}^{n} r_\ell^1 \left(\frac{\partial f}{\partial x_\ell}\right)_0 = 0. \qquad (5.52)$$

The strategem used to identify r_ℓ^1 is to relate it to $\bar{r}_\ell^0$ and $\left(\frac{\partial f}{\partial x_\ell}\right)_1$ as follows:

$$r_\ell^1 = \left(\frac{\partial f}{\partial x_\ell}\right)_1 + \theta_1 r_\ell^0 \qquad \ell = 1, 2, \ldots, n. \qquad (5.53)$$

The constant θ_1 is chosen to satisfy (5.52). Substitution of this assumed form for $\bar{r}_\ell^1$ into (5.52) gives

$$\sum_{\ell=1}^{n} \left(\frac{\partial f}{\partial x_\ell}\right)_1^2 + \theta_1 \sum_{\ell=1}^{n} \left(\frac{\partial f}{\partial x_\ell}\right)_1 r_\ell^0 - \sum_{\ell=1}^{n} \left(\frac{\partial f}{\partial x_\ell}\right)_1 \left(\frac{\partial f}{\partial x_\ell}\right)_0$$
$$- \theta_1 \sum_{\ell=1}^{n} \left(\frac{\partial f}{\partial x_\ell}\right)_0^2 = 0.$$

Because the point $\bar{x}^1$ was chosen to maximize f in the direction $\bar{r}^0$, which was equal to the gradient direction, the middle two terms in this equation vanish by virtue of (5.42). Hence θ_1 is obtained directly

$$\theta_1 = \frac{\sum_{\ell=1}^{n} \left(\dfrac{\partial f}{\partial x_\ell}\right)_1^2}{\sum_{\ell=1}^{n} \left(\dfrac{\partial f}{\partial x_\ell}\right)_0^2} . \tag{5.54}$$

Since θ_1 is readily computed from the information available at the two test points, it is thus possible to select $\bar{r}^1$ by (5.53) with the assurance that it is conjugate to $\bar{r}^0$.

Continuation of the search in the direction $\bar{r}^1$ locates the point $\bar{x}^2$ by the maximizing procedure. Again the gradient of f is evaluated at $\bar{x}^2$, and a conjugate direction for further exploration is sought. This direction is obtained as before. The partial derivative of f with respect to x_ℓ at $\bar{x}^2$ is written as

$$\left(\frac{\partial f}{\partial x_\ell}\right)_2 = c_\ell - \sum_{i=1}^{n} q_{\ell_i} x_i^0 - d_0^* \sum_{i=1}^{n} q_{\ell_i} r_i^0 - d_1^* \sum_{i=1}^{n} q_{\ell_i} r_i^1.$$

Subtraction of the same derivative evaluated at $\bar{x}^0$ yields

$$\left(\frac{\partial f}{\partial x_\ell}\right)_2 - \left(\frac{\partial f}{\partial x_\ell}\right)_0 = - d_0^* \sum_{i=1}^{n} q_{\ell_i} r_i^0 - d_1^* \sum_{i=1}^{n} q_{\ell_i} r_1^1 .$$

Multiplication of this equation by r_ℓ^2 and summation from $\ell = 1$ to $\ell = n$ gives

$$\sum_{\ell=1}^{n} r_\ell^2 \left(\frac{\partial f}{\partial x_\ell}\right)_2 - \sum_{\ell=1}^{n} r_\ell^2 \left(\frac{\partial f}{\partial x_\ell}\right)_0 =$$

$$- d_0^* \sum_{\ell=1}^{n} \sum_{i=1}^{n} q_{\varrho_i} r_i^0 r_\varrho^2 - d_1^* \sum_{\ell=1}^{n} \sum_{i=1}^{n} q_{\varrho_i} r_i^1 r_\varrho^2.$$

For $\bar{r}^2$ to be conjugate to $\bar{r}^0$ and $\bar{r}^1$, the right-hand term of this equation must vanish and

$$\sum_{\ell=1}^{n} r_\varrho^2 \left(\frac{\partial f}{\partial x_\varrho}\right)_2 - \sum_{\ell=1}^{n} r_\varrho^2 \left(\frac{\partial f}{\partial x_\varrho}\right)_0 = 0. \quad (5.55)$$

Capitalizing on the successful hypothesis for $\bar{r}^1$, we assume the following form for $\bar{r}^2$

$$r_\varrho^2 = \left(\frac{\partial f}{\partial x_\varrho}\right)_2 + \theta_2 r_\varrho^1 \qquad \ell = 1, 2, \ldots, n \quad (5.56)$$

where θ_2 is a constant to be determined.

Substitution of this expression for r_ℓ^2 into the first term of (5.55) gives

$$\sum_{\ell=1}^{n} \left(\frac{\partial f}{\partial x_\varrho}\right)_2 \left[\left(\frac{\partial f}{\partial x_\varrho}\right)_2 + \theta_2 r_\varrho^1\right] = \sum_{\ell=1}^{n} \left[\left(\frac{\partial f}{\partial x_\varrho}\right)_2\right]^2 \quad (5.57)$$

by virtue of (5.42).

In order to reduce the second term, r_ℓ^2 is first written as

$$r_\varrho^2 = \left(\frac{\partial f}{\partial x_\varrho}\right)_2 + \theta_2 \left[\left(\frac{\partial f}{\partial x_\varrho}\right)_1 + \theta_1 r_\varrho^0\right]$$

where r_ℓ^1 has been eliminated by means of (5.53). Finally, it is noted that $\left(\frac{\partial f}{\partial x_\varrho}\right)_2$ is equivalent to the following

$$\left(\frac{\partial f}{\partial x_\ell}\right)_2 = \left(\frac{\partial f}{\partial x_\ell}\right)_1 - d_1^* \sum_{i=1}^{n} q_{\ell_i} r_i^1$$

so that the final form of r_ℓ^2 is

$$r_\ell^2 = (\theta_2 + 1)\left(\frac{\partial f}{\partial x_\ell}\right)_1 - d_1^* \sum_{i=1}^{n} q_{\ell_i} r_i^1 + \theta_1 \theta_2 r_\ell^0 .$$

Thus the second term of (5.55) becomes

$$\sum_{\ell=1}^{n} \left(\frac{\partial f}{\partial x_\ell}\right)_0 r_\ell^2 = (\theta_2 + 1) \sum_{\ell=1}^{n} \left(\frac{\partial f}{\partial x_\ell}\right)_0 \left(\frac{\partial f}{\partial x_\ell}\right)_1$$

$$- d_1^* \sum_{\ell=1}^{n} \sum_{i=1}^{n} q_{\ell_i} r_i^1 \left(\frac{\partial f}{\partial x_\ell}\right)_0 + \theta_2 \theta_1 \sum_{\ell=1}^{n} \left(\frac{\partial f}{\partial x_\ell}\right)_0 r_\ell^0 .$$

Because $r_\ell^0 = \left(\frac{\partial f}{\partial x_\ell}\right)_0$, the first term on the right vanishes by virtue of (5.42). The second term vanishes because $\bar{r}^1$ and $\bar{r}^2$ are conjugate directions. Substitution for θ_1 from (5.54) into the third terms yields the following

$$\sum_{\ell=1}^{n} \left(\frac{\partial f}{\partial x_\ell}\right)_0 r_\ell^2 = \theta_2 \sum_{\ell=1}^{n} \left[\left(\frac{\partial f}{\partial x_\ell}\right)_1\right]^2 \tag{5.58}$$

Substitution of (5.57) and (5.58) into (5.55) gives

$$\theta_2 = \frac{\sum_{\ell=1}^{n} \left[\left(\frac{\partial f}{\partial x_\ell}\right)_2\right]^2}{\sum_{\ell=1}^{n} \left[\left(\frac{\partial f}{\partial x_\ell}\right)_1\right]^2} . \tag{5.59}$$

Since θ_2, like θ_1, is readily computed from the information on the partial derivatives of the objective function at the points $\bar{x}^1$ and $\bar{x}^2$, the direction $\bar{r}^2$ is easily specified.

Arguments similar to those already used show that, at the kth test point, the direction $\bar{r}^k$ is conjugate to all previous directions if

$$r_{\ell}^k = \left(\frac{\partial f}{\partial x_{\ell}} \right)_k + \theta_k r_{\ell}^{k-1} \qquad \ell = 1, 2, \ldots, n \tag{5.60}$$

where

$$\theta_k = \frac{\displaystyle\sum_{\ell=1}^{n} \left[\left(\frac{\partial f}{\partial x_{\ell}} \right)_k \right]^2}{\displaystyle\sum_{\ell=1}^{n} \left[\left(\frac{\partial f}{\partial x_{\ell}} \right)_{k-1} \right]^2} . \tag{5.61}$$

Experience with the conjugate gradient technique has been very good, particularly in the vicinity of the optimum where most functions exhibit a quadratic behavior. When the function to be maximized is not quadratic to begin with, the algorithm will not have converged after the n iterations. The recommended practice in this case is to reinitiate the algorithm at the nth point.

The technique is easily programmed for a digital computer, and several existing codes are available. A number of variants of the scheme described here have been proposed, and in most cases they achieve comparable efficiency.

Example 5.5 Resolve the problem of examples 5.3 and 5.4 one more time, using the gradient values to obtain the conjugate directions.

Solution The problem is to maximize

$$z = f = x_1 + x_2 - \frac{1}{2} (x_1^2 + 2x_1 x_2 + 2x_2^2).$$

We begin the search at (1,1) where

$$\left(\frac{\partial f}{\partial x_1} \right)_0 = 1 - x_1 - x_2 = 1 - 1 - 1 = -1$$

$$\left(\frac{\partial f}{\partial x_2} \right)_0 = 1 - x_1 - 2x_2 = 1 - 1 - 2 = -2.$$

The first search is in the gradient direction, and it was shown in example 5.3 that this terminates at $x_1^1 = 8/13$, $x_2^1 = 2/13$. Here

$$\left(\frac{\partial f}{\partial x_1} \right)_1 = 1 - 8/13 - 3/13 = 2/13$$

$$\left(\frac{\partial f}{\partial x_2} \right)_1 = 1 - 8/13 - 6/13 = -1/13.$$

Hence

$$r_1^1 = 2/13 + \frac{\left(\frac{2}{13} \right)^2 + \left(\frac{1}{13} \right)^2}{\left(-1 \right)^2 + \left(-2 \right)^2} \; (-1) = \frac{25}{169}$$

$$r_2^1 = -1/13 + \frac{\left(\frac{2}{13} \right)^2 + \left(-\frac{1}{13} \right)^2}{\left(-1 \right)^2 + \left(-2 \right)^2} \; (-2) = \frac{-15}{169}.$$

Along the search path, x_1 and x_2 have the values

$$x_1 = \frac{8}{13} + \frac{25}{169} d_1$$

$$x_2 = \frac{3}{13} - \frac{15}{169} d_1$$

Hence the objective function is

$$z = \frac{78}{169} + \frac{65}{(13)^3} \, d_1 - \frac{1}{2} \, \frac{325}{(13)^4} \, d_1^2$$

and

$$d_1^* = \left(\frac{13}{5} \right).$$

This gives $x_1^2 = 1$, $x_2^2 = 0$, which is the optimizing point.

5.3.2 Handling Constraints The presence of inequality constraints complicates the use of the conjugate direction technique. However, the method can be used in a constrained problem as long as the test points are in the interior of the feasible region. When a test point on the boundary is reached, the current cycle of conjugate directions is terminated. One of the techniques for obtaining constrained ascent directions discussed in the last section can then be invoked and the search continued in this way. If the search direction again points into the feasible region, a new cycle of conjugate directions can be initiated.

Bibliography

General sources of methods and references on indirect optimization techniques, in addition to those already cited are

1. Beveridge, Gordon S. and Robert S. Schechter. *Optimization: Theory and Practice.* McGraw Hill Book Co., New York, 1970.

2. Pun, Lucas. *Introduction to Optimization Practice.* John Wiley & Sons, Inc., New York, 1969.

3. Rosenbrock, H.H. and C. Storey. *Computational techniques for chemical engineers.* Pergamon Press, London, 1966.

The gradient projection method was proposed by Rosen:

4. Rosen, J.B. The gradient projection method for nonlinear programming: Part I. Linear constraints. *J. Soc. Ind. Appl. Math.*, 8, pp. 181—217, 1960.

5. Rosen, J.B. The gradient projection method for nonlinear programming: Part II. Nonlinear constraints. *J. Soc. Ind. Appl. Math.*, 9, pp. 514—532, 1961.

Alternate approaches to determining directions of constrained ascent are given by Zoutendijk:

6. Zoutendijk, G., *Methods of Feasible Directions.* Elsevier, Amsterdam, 1960.

Comprehensive discussions of the conjugate-direction approach are given in

7.   Fletcher, R. and M.J.D. Powell. A rapidly convergent descent method of minimization. *Comput, J.* 6, pp. 163—168, 1963.

8.   Fletcher, R. and C.M. Reeves. Function minimization by conjugate directions. *Comput. J.* 7, pp. 149—154, 1964.

Several techniques which attempt to compromise between first and second order methods have been proposed. A particularly effective one is due to Marquardt:

9.   Marquardt, D.W. An Algorithm for Least Squares Estimation of Nonlinear Parmaeters. *J. Soc. Ind. Appl. Math.* 11, pp. 431, 1963.

Finally, several perceptive reviews of the present status of parameter optimization routines are collected in

10.   Fletcher, R., ed., *Optimization.* Academic Press, London, 1969.

11.   Hadley, G. *Nonlinear and Dynamic Programming.* Addison-Wesley Publishing Co., Inc., Reading, Massachusetts, 1964.

6
Discrete Dynamic Programming

6.1 INTRODUCTION

Many problems in resource allocation, engineering and operations research have a particular structure which allows their optimization to be carried out in a much more efficient manner than by the more general techniques of multivariable search or nonlinear programming. One class of such problems are those which have a stagewise character in which the independent variables appear in a sequential fashion in the objective function. These problems are amenable to a solution procedure devised by Richard Bellman in the early 1950s (3,4). Further developments of the technique, which is known as dynamic programming, have been carried out since then by Bellman and his associates at the RAND Corporation (5) and many engineering applications have been reported (1,2,7).

To illustrate the nature of stagewise processes and indicate how dynamic programming can be used to solve optimization problems associated with such processes, the following simple example is considered.

6.2 AN EXAMPLE OF A STAGEWISE PROCESS

Extraction is a chemical engineering operation whereby a compound or compounds in liquid solution with one or more other compounds is removed from the solution by contacting it with a second liquid known as an extracting solvent. The compound in question dissolves in the extracting solvent, which is then removed by physical means from contact with the original liquid.

There are a number of ways in which this process is carried out in practice. Figure 6.1 depicts one such way, whereby N fresh streams of solvent are added at N individual locations in a cross-flow manner to the stream of feed solution. Such an arrangement is known as cross-current extraction. Each of the N contact points is considered to be an ideal stage in which the mixing between feed and solvent is accomplished and then separation of streams is effected.

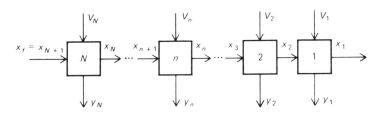

Figure 6.1

The solute is assumed to equilibrate between the two streams.

A solute material balance around the nth stage is:

$$q(x_{n+1} - x_n) = W_n y_n \qquad n = 1, 2, \ldots, N \tag{6.1}$$

where q = flow rate of feed solution,

x = concentration of solute in raffinate,

y = concentration of solute in extract,

W_n = flow rate of solvent to n^{th} stage, and

x_{N+1} = feed concentration of solute.

Let $V_n = W_n/q$, a dimensionless flow term, so that (6.1) becomes

$$x_{n+1} - x_n = V_n \, y_n \qquad n = 1, \, 2, \, \ldots, \, N.$$
$$(6.2)$$

The equilibrium expression relating y to x may be expressed generally by stating $y = y(x)$. Later it will be convenient to use a specific form for this relationship.

Equations (6.2) together with the equilibrium relationship constitute a set of equations which describe the behavior of the cross-current extraction unit. They are a model of the system. In order to frame an optimization problem for this unit, it is necessary, in addition to the system model, to postulate a performance criterion or objective function which, if extremized, would signify optimal performance.

It is possible to suggest several performance criteria for the extraction system; one often used is operating profit. For a typical stage, say the nth stage, the operating profit P_n may be considered as the value of the material extracted less the cost of solvent used. Quantitatively, this is expressed in the following way:

$$P_n = P_e \, [W_n \cdot y_n] \, - \, C_w \, W_n \qquad (6.3)$$

where P_e is the price obtained per unit of extracted material and C_w is the cost per unit of solvent. Equation (6.3) may be normalized by dividing through by $P_e \cdot q$. Define a normalized return $r_n = P_n/P_e q$ and a relative cost $\lambda = C_W/P_e$; then (6.3) becomes

$$r_n = V_n \, (y_n - \lambda). \qquad (6.4)$$

The normalized return for the entire N-stage unit, R, is the sum of the individual stage returns

$$R = \sum_{n=1}^{N} r_n$$

and it is R that we wish to maximize.

Mathematically, R is a function of the $2N$ variables $V_1, V_2, \ldots, V_N$ and $x_1, x_2, \ldots, x_N$. However, there are N subsidiary equations or constraints (6.1) relating these variables so that only N selections or decisions need be made to optimize R. A logical distinction can be made between the two sets of variables on which R depends. The composition variables, $x_1, x_2, \ldots, x_N$, characterize the amount of solute in the process stream at each stage. We will find it convenient to eliminate these variables by means of the constraint equations, in a manner reminiscent of that used in section 1.5. Hence it is appropriate to refer to the x-variables as state variables.

The variables $V_1, V_2, \ldots, V_N$, representing the normalized solvent rates to the various stages, on the other hand, represent those quantities which physically would be manipulated to exercise control over the extraction unit. They are also the variables which will be selected by the dynamic programming technique. Hence these variables are called decision variables.

With this distinction between state and decision variables having thus been drawn, it is useful to consider the decision variables in our optimization problem as the independent variables, with the state variables replaced in the objective function by means of the constraint or system equations, as we did in section 1.5. This point will be developed more formally in a later section, although it will be illustrated in the development of a solution to the cross-current extraction optimization problem.

A Solution Procedure If N were equal to one — i.e., if we had a single stage process —

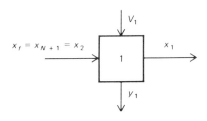

Figure 6.2

Then the optimization problem would be:

$$\text{maximize} \quad R = r_1 = V_1 \ (y_1 - \lambda) \quad\quad (6.5)$$

$$\text{subject to} \quad x_f - x_1 = V_1 \ y_1 \quad\quad (6.6)$$

$$\text{and} \quad\quad y_1 = y(x_1). \quad\quad (6.7)$$

Although three variables are involved, x_1 and y_1 can conceivably be solved for in terms of V_1 by solving (6.6) and (6.7), thereby reducing R to a function of V_1 alone. In this case, any of several common optimum seeking methods can be employed. To illustrate this idea more specifically, consider a simple form of the equilibrium function — namely a linear relationship,

$$y_1 = \alpha x_1, \quad\quad (6.8)$$

so that, upon substitution of (6.8) into (6.6), the following equation results:

$$\left(V_1 + \frac{1}{\alpha} \right) y_1 = x_f \quad\quad (6.9)$$

which, upon substitution into R, reduces the system profit to a function of V_1 alone:

$$R = V_1 \left[\frac{x_f}{\left(V_1 + \dfrac{1}{\alpha} \right)} - \lambda \right]. \quad\quad (6.10)$$

One can easily verify that R is maximized for

$$V_1 = \left(\frac{x_f}{\alpha \lambda} \right)^{\frac{1}{2}} - \frac{1}{\alpha}. \quad\quad (6.11)$$

Suppose now that, instead of a single stage, the process consisted of two stages. Again consider a linear equilibrium (6.8) to prevail.

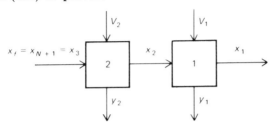

Figure 6.3

Repetition of the above procedure would reduce the objective function R to a dependence on V_1 and V_2, and the extremizing values of these variables would be those which reduced the gradient of R to zero. This traditional approach makes the N decision (in this case, $N = 2$) simultaneously. If a procedure for making the decisions sequentially could be developed, whereby the N decisions were made one at a time, it is not difficult to appreciate the reduction in problem complexity that would accompany the technique.

It can be argued that, in a stagewise process such as depicted in Figure 6.3, the decision made on the last stage (numbered here as 1, since, as we shall see, it will be the first stage to be looked at) should not influence any of the preceding stages. One might argue further that the decision made on this stage should be identical to that made in the one-stage process — the only difference being that here the inlet stream composition, x_2, is no longer the feed composition. As a matter of fact, this composition is directly influenced by the value assigned to V_2 and will be a result of the optimization procedure specifying this value. However, to continue the original argument, it is easily appreciated and can be shown that, regardless of what value x_2 takes on as a result of the optimization procedure, V_1 will have to be selected to maximize the return remaining for the process. Optimization is confined to stage one; x_2 is treated as a given quantity, much as x_f was treated as preassigned whenever the process consisted of a single stage. But note carefully that, since x_2 is not known beforehand (as x_f was in the single stage case), a single optimizing value of V_1 cannot be at once specified. Indeed the choice of V_1 depends upon x_2, and hence the exact optimizing value will only be known when V_2 is also selected, thereby specifying x_2. However, the decision on the number-one stage can be made separately — regarding x_2 as a given item. Therein lies the secret of the sequential decision process, "which does the best it can with whatever it has, for if it does not at least do this, it can never hope to attain truly optimum performance" (2).

This single idea forms the crux of dynamic programming and is elegantly phased in the discipline's primary postulate, the principle of optimality: "An optimal policy has the property that whatever the initial state and initial decision are, the remaining decisions must constitute an optimal policy, with regard to the state resulting from the first decision" (3). Thus, for our purpose, we may look on the state x_f as an initial state. The principle of optimality states that, whatever V_2 is, the remaining decisions (in this case, V_1) must constitute an optimal policy (max R for the remaining stages — here $R = r_1$).

For a three-stage process, the decisions on stage number 1 are made analogously to those for the two-stage process. We might symbolize the results by a plot of x_n versus n (Figure 6.4). For a number of values of x_2 (depicted as circles), an optimum value of V_1 is selected. This may be symbolized as $V_1^* (x_2)$ the $*$ denoting optimum and the functional dependence to remind us that the optimum value depends upon x_2. This value of V_1 causes x_1 to take on a unique value (6.6) depicted as a square, connected to the circle representing x_2. The connecting line depicts a path the x-variable follows in proceeding from x_2 to x_1. Moreover, if the process should ever be carried out in such a manner, regardless of the number of stages, that it arrives at any one of the circled values of x_2, the optimal transition from x_2 to x_1 will be along the appropriate line and V_1 will take on the

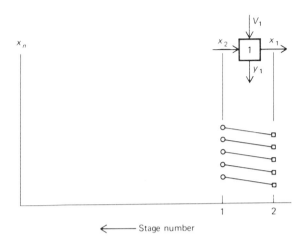

Figure 6.4

value $V_1^*(x_2)$ corresponding to the x_2 value.

Consider now the three-stage process to be composed of an initial stage, stage 3, and a two-stage process comprised of stages 1 and 2. Regardless of what goes on in stage 3 — i.e., irrespective of the value x_3, — the decisions V_1 and V_2 must be made in such a way to maximize R for the two-stage process, $R = r_1 + r_2$. However, this is identical to the problem of optimizing a two-stage process except that now the variable x_3 is not specified. To ensure that we examine the process for the condition resulting from an optimal decision on stage 3, we must locate optimal paths for a number of values of x_3. Suppose that, for the moment, our attention is focused on a particular value of x_3, depicted on Figure 6.5 as a triangle.

We know that, if we select V_2 in such a manner as to connect the triangle with any of the circled values of x_2, the optimal policy is determined from there. Thus we might depict 5 possible routes from the value of x_3 at the triangle out to the end of the process by selecting 5 values of V_2 (V_2^1, V_2^2, V_2^3, V_2^4, V_2^5) which connect the $\triangle$ value of x_3 to the 0 values of x_2. We can then compute the return for each path between $\triangle$ and $\square$. That path which maximizes the 2-stage return will be the optimal path

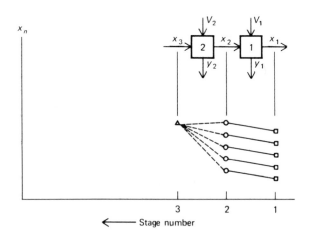

Figure 6.5

from the △ value of x_3 to the corresponding value at
stage 1. This path will have associated with it values of
both V_1 and V_2 which are optimal with respect to the
particular value of x_3; i.e., they constitute an optimal policy dependent upon x_3. We might symbolize them now
as $[V_1^*(x_3), V_2^*(x_3)]$.

In actual practice, the selection of a particular value
for V_2 may result in an x_2 value other than depicted by
one of the 0 values. In this case, it would be necessary to
obtain V_1 and the one-stage return by interpolating between adjacent x_2 values for which an optimal policy is
known.

If a number of values of x_3 are examined, the above
procedure can be repeated for each, and a family of paths
of x_n versus n constructed, with a particular policy,
$[V_1^*(x_3), V_2^*(x_3)]$ associated with each. One of these
paths will ultimately turn out to be optimal with respect
to the given inlet feed concentration. For the moment,
this path is not known; it is instead imbedded in the family of paths which have x_3 as a parameter. In general,
more than five paths would be established; the use of five
paths in Figures 6.4-6.6 is simply to illustrate the method.

Suppose now we have repeated this procedure $(N-1)$ times and determined optimal policies $[V_1^* (x_N), \ V_2^* (x_N), \ \ldots, \ V_{N-1}^* (x_N)]$ as functions of x_N which characterize optimal paths of x_n as a function of n for various values of x_N, Figure 6.6.

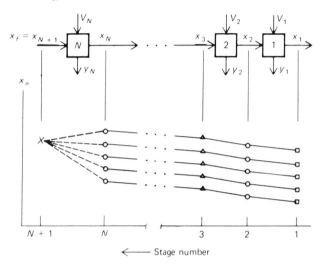

Figure 6.6

As we prepare to add another stage, following the established format, we notice that $x_{N+1} = x_f$. Unlike previous cases, x_{N+1} is not the result of a previous decision but rather has a definite, preassigned value, x_f. Thus we need only consider the value of $x_{N+1} = x_f$, represented by an X on the diagram. Connect the X to the 0's by means of choosing V_N. From 0 to $\square$, for each 0 value of x_N, the optimal path and policy are known so that the computation of R can be accomplished:

$$R = \sum_{n=1}^{N} r_n.$$

That value of V_N is selected which connects the X point representing x_f to the 0 point representing x_N such that the policy consisting of this V_N value and the $N-1$ stage, optimal policy, $[V_1^* (x_N), V_2^* (x_N), \ldots, V_{N-1}^*(x_N)]$ associated with the 0 maximize R. This completes the problem solution; the optimal policy $[V_1^* (x_f), V_2^* (x_f), \ldots, V^* (x_f)]$ is the N-stage policy which maximizes

$$R = \sum_{n=1}^{N} r_n.$$

6.3 A FORMAL SOLUTION ALGORITHM

The problem posed in the last section may be formalized mathematically in the following way. Recall that the decision on stage 1 was first made by selecting V_1 to maximize r_1 with x_2 treated as a constant. In general, r_1 may be a function both of V_1 and x_2 by the system or transformation equation. Symbolically, we may identify this relationship for any stage as follows

$$x_n = T(V_n, x_{n+1}) \qquad n = 1, 2, \ldots, N \tag{6.12}$$

$$r_n = r_n(x_n, V_n) \qquad n = 1, 2, \ldots, N. \tag{6.13}$$

In order to maximize r_1 with x_2 fixed, (6.12) must be substituted into equation (6.13). Thus we represent the process mathematically as

$$\underset{\{V_1\}}{\text{Max}} \left\{ r_1 \left[T(V_1, x_2), V_1 \right] \right\}. \tag{6.14}$$

This is to be read as follows: maximize the expression $r_1 [T (V_1, x_2), V_1]$ by selecting V_1 and considering x_2 as constant. The resultant value of r_1, calculated by using the maximizing value of V_1, is a function, therefore, of x_2 and may be represented by $f_1 (x_2)$:

$$f_1(x_2) = \underset{\{V_1\}}{\text{Max}} \ \left\{ r_1(T[V_1, \ x_2], \ V_1) \right\} \ . \quad (6.15)$$

For a two-stage process, x_3 is considered fixed. We use $f_2(x_3)$ as the maximum of $r_1 + r_2$ with respect to V_1 and V_2:

$$f_2(x_3) = \underset{\{V_1, \ V_2\}}{\text{Max}} \ \left\{ r_2 \ [x_2, \ V_2] + r_1 \ [x_1, \ V_1] \right\} \ . \quad (6.16)$$

Note that, if V_2 is selected by the maximizing procedure, this fixes x_2 by (6.12) since x_3 is considered as fixed. However, with x_2 fixed, the maximum of r_1 has already been found; it is given by $f_1(x_2)$. Hence there is no need to look further for a maximizing value of V_1, and (6.16) may be written as follows:

$$f_2(x_3) = \underset{\{V_2\}}{\text{Max}} \ \left\{ r_2 \ [T(V_2, \ x_3), \ V_2] + f_1 \ [T(V_2, \ x_3)] \right\} \ . \quad (6.17)$$

In general, for the nth stage,

$$f_n(x_{n+1}) = \underset{\{V_n\}}{\text{Max}} \ \left\{ r_n \ [T(V_n, \ x_{n+1}), \ V_n] \right.$$
$$\left. + f_{n-1} \ [T(V_n, \ x_{n+1})] \right\} \ . \quad (6.18)$$

This equation is to be interpreted in the following way. Select V_n to maximize the curved-bracketed term, which is equal to $r_1 + r_2 + \ldots + r_n$, while treating x_{n+1} as constant. Under these conditions, the curved-bracketed term depends upon V_n in two ways; the first dependence is that of r_n upon V_n and the second comes about by realizing that, with x_{n+1} fixed, the choice of V_n identifies x_n, for which values of $f_{n-1}(x_n)$ have been tabulated by the previous or $n-1^{th}$ such process. The value of $f_0(x_1) = 0$ initiates the solution procedure, which repeats the functional equation (6.18) N times.

When dynamic programming is used to solve a stage-wise optimization process, provision must be made to store values of $f_{n-1}(x_n)$. The optimization of the term $(r_n + f_{n-1})$ can be carried out by the techniques discussed in the previous chapters. Finally, it is to be realized that stagewise processes with several state and control variables can be handled by the method.

The computational efficiency of dynamic programming derives from the fact that at each step all unfavorable combinations of variables are eliminated and not carried over to the next step where they would have to be considered. Thus, one would expect the technique to provide advantage where a large number of stages are involved — and this turns out to be the case. The difficulty associated with dynamic programming is that, whenever a moderate number of state variables are included in the problem statement, the storage required for the various values of the f's becomes excessive. Bellman refers to this unhappy situation as "the curse of dimensionality", and we can do little to improve on his descriptive terminology.

We will have more to say on this subject in a later section. At the moment, however, it is useful to apply the general solution procedure to the cross-current extraction problem.

6.4 APPLICATION OF FUNCTIONAL EQUATION TO CROSS-CURRENT EXTRACTION UNIT WITH LINEAR EQUILIBRIUM

Consider again the cross-current extraction unit for the case of a linear equilibrium. Dynamic programming begins by considering $f_1(x_2)$:

$$f_1(x_2) = \underset{\{V_1\}}{\text{Max}} \{[r_1] = \underset{\{V_1\}}{\text{Max}} [V_1(\alpha x_1 - \lambda)] $$

$$= \underset{\{V_1\}}{\text{Max}} [V_1\left(\frac{\alpha x_2}{1 + \alpha V_1} - \lambda\right)]. \qquad (6.18)$$

Notice that the quantity to be maximized has been expressed in terms of V_1 and x_2. The value of V_1, V_1^*. which maximizes the square bracketed term is readily found by setting $\partial[\ ]/\partial V_1 = 0$

$$V_1^* = \sqrt{\frac{x_2}{\alpha\lambda}} - \frac{1}{\alpha}. \qquad (6.19)$$

If the value of V_1 is substituted into (6.18), an explicit representation of $f_1(x_2)$ results:

$$f_1(x_2) = x_2 - \sqrt{\frac{\lambda}{\alpha}}. \qquad (6.20)$$

The recursive functional equation of dynamic programming yields for the two-stage process the following:

$$f_2(x_3) = \underset{\{V_2\}}{\text{Max}} [r_2 + f_1(x_2)]$$

$$f_2(x_3) = \underset{\{V_2\}}{\text{Max}} \left[V_2 \left(\frac{\alpha x_3}{1 + \alpha V_2} - \lambda \right) \right.$$

$$\left. + \frac{x_3}{1 + \alpha V_2} - \sqrt{\frac{\lambda}{\alpha}} \right]. \qquad (6.21)$$

It can be shown that V_2^*, the value of V_2 which maximizes the square bracketed term of (6.21), is equal to V_1^*. Upon repetition of this process, it can be further shown that $V^* = V_1^* = V_2^* = \ldots = V_n^* = \ldots = V_N^*$. That is, the optimal policy is to split the solvent flow equally to each stage; the amount of this flow is obtained by applying the system equation $N+1$ times, to give

$$\alpha x_{N+1} = \lambda(1 + \alpha V^*)^{N+1}. \qquad (6.22)$$

However, since $x_{N+1} = x_f$, the feed composition, it follows that

$$V^* = \frac{1}{\alpha}\left[\left(\frac{\alpha x_f}{\lambda}\right)^{\frac{1}{N+1}} - 1\right] \qquad (6.23)$$

which is the solution we seek.

6.5 RELATIONSHIP OF DYNAMIC PROGRAMMING TO NONLINEAR PROGRAMMING

Suppose we postulate a nonlinear programming problem as follows:

$$\text{Max } R = \sum_{n=1}^{N} r_n\,(V_n) \qquad (6.24)$$

subject to

$$\sum_{n=1}^{N} g_n\,(V_n) = k. \qquad (6.25)$$

Here the objective function, R is a function of the N independent variables, $V_1, V_2, \ldots, V_N$, subject to the constraint equation (6.25). Thus, only $N-1$ of the N variables are truly independent.

Various other types of constraints can be imposed on the system of equations, such as requiring that

$$V_n \geqslant 0 \qquad n = 1, 2, \ldots, N \qquad (6.26)$$

and/or

$$g_n(V_n) \geqslant 0 \qquad n = 1, 2, \ldots, N. \qquad (6.27)$$

Now we define

$$x_{n+1} - x_n = g_n(V_n) \tag{6.28}$$

and require that $\qquad x_{N+1} = k$

$$x_1 = 0. \tag{6.29}$$

Then the constraint equation (6.25) is automatically satisfied, and the problem has been reduced to one to which the dynamic programming technique directly applies.

If we have several constraints of the type (6.25), we can incorporate them by introducing a state variable for each such constraint. However, it should be recalled that many state variables correspond to a cursed problem, i.e., they require vast amounts of storage capacity. We can be saved from this unfortunate dilemma by employing the Lagrange multiplier concept. In order to be specific, consider ˙that we have M constraints.

$$\sum_{n-1}^{N} g_{n,i}(V_n) = b_i \qquad i-1, 2, \ldots, M \tag{6.30}$$

If we incorporate M' of these $(M' \le M)$ into the objective function by means of M' Lagrange multipliers, λ_1, λ_2, . . . , $\lambda_{M'}$, a new objective function evolves:

$$\underset{\{V_n\}}{\text{Max}} \left\{ \sum_{n=1}^{N} r_n(V_n) + \sum_{i=1}^{M'} \lambda_i \left[b_i - \sum_{n=1}^{N} g_{n,i}(V_n) \right] \right\}.$$

The remaining $(M-M')$ constraints still apply and are handled by introducing $(M-M')$ state variables.

The problem is then solved by specifying a set of values for the Lagrange Multipliers, $\lambda_1, \lambda_2, \ldots, \lambda_{M'}$, and applying the dynamic programming algorithm to the resultant problem. If the set of values were chosen correctly, all of the M' constraints would be satisfied by the optimal policy just computed. If they were not satisfied, a new set of values would have to be assigned to the multipliers and the process repeated until the constraints were satisfied.

Thus, this method involves solving a simpler problem many times. It is a trade off between time of solution and computer memory. A similar result is obtained if penalty functions are used.

6.6 EXTENSIONS TO NONSERIAL SYSTEMS

The applications of dynamic programming presented so far here have been for systems exhibiting a serial structure. The nth stage return depended only upon the nth stage output and decision variables while the nth stage output depended only upon the input and decision variables for that stage. Many stagewise processes of practical interest have more complex structures. In most cases, the functional equation of dynamic programming can be modified to handle the increased complexity.

As an example of a nonserial system, consider the cross-current extraction system modified by the presence of a recycle stream (Figure 6.7)

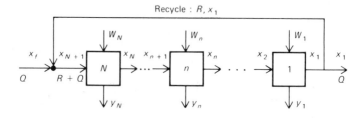

Figure 6.7

The solute material balance about the nth stage is

$$(Q + R)(x_{n+1} - x_n) = w_n y_n \qquad (6.31)$$

where Q = flow rate of feed solution,

R = flow rate of recycle solution

x = concentration of solute in raffinate,

y = concentration of solute in extract, and

w_n = flow rate of solvent to n^{th} stage.

Let $V_n = W_n/(Q + R)$. Equation (6.31) becomes

$$x_{n+1} - x_n = V_n y_n. \tag{6.32}$$

The material balance on the recycle stream is

$$Qx_f + Rx_1 = (Q + R) x_{N+1}.$$

Let $q = Q/(Q + R)$ and $r = R/(Q + R)$;

then
$$qx_f + rx_1 = x_{N+1}. \tag{6.33}$$

The performance index is as before:

$$z = \sum_{n=1}^{N} V_n(y_n - \lambda) \tag{6.34}$$

and, for convenience, the linear equilibrium is assumed

$$y_n = \alpha x_n.$$

The solution procedure for this cyclic system is to consider the N-decision process to consist of $N-1$ optimal decisions, V_2^*, V_3^*, . . . , V_N^* and an optimum value assigned to x_1 : x_1^*. The dynamic programming algorithm is applied to the system with x_1 fixed at some value and the optimum allocations of extracting solvent determined for stages 2 through N for this value of x_1. This strategy yields the return for the process as a function of x_1. The procedure is then repeated for various x_1 until the maximum value is located for the objective function.

To make these ideas quantitative, define a variable S_n as

$$S_n = \sum_{i=1}^{n} V_i(\alpha x_i - \lambda). \tag{6.35}$$

A further variable f_n is defined as

$$f_n(x_1, x_{n+1}) = \underset{\{V_2, V_3, \ldots, V_n\}}{\text{Max}} [S_n(x_1, V_2, \ldots, V_n, x_{n+1})]. \tag{6.36}$$

Thus f_n is the maximum return from an n-stage process, with x_1 and x_{n+1} treated as constants. f_0 is zero, and the principle of optimality gives

$$f_n(x_1, x_{n+1}) = \underset{\{V_n\}}{\text{Max}} [V_n(\alpha x_n - \lambda) + f_{n-1}(x_1, x_n)]. \tag{6.37}$$

The maximum cyclic return, S_N^*, is obtained by selecting x_1 so that

$$S_N^* = \underset{\{x_1\}}{\text{Max}} [f_N(x_1, x_{N+1})]. \tag{6.38}$$

For the cross-current extraction unit, the dynamic programming algorithm begins by evaluating $f_1(x_1, x_2)$. With x_1 and x_2 treated as fixed, no optimization can be made on stage 1. V_1 is fixed as

$$V_1 = \frac{1}{\alpha}\left(\frac{x_2}{x_1} - 1\right). \tag{6.39}$$

Therefore,

$$f_1(x_1,\ x_2) = \frac{1}{\alpha}\left(\frac{x_2}{x_1} - 1\right)\left(\alpha x_1 - \lambda\right).$$

For the two-stage process, x_1 and x_3 are held constant. The variable x_2 is replaced by

$$x_2 = \frac{x_3}{1 + \alpha V_2}$$

so that

$$f_2(x_1,\ x_3) = \underset{V_2}{\text{Max}}\ \left\{V_2\left(\frac{\alpha x_3}{1 + \alpha V_2} - \lambda\right)\right.$$

$$\left. + \frac{1}{\alpha}\left(\frac{x_3}{x_1(1 + \alpha V_3)} - 1\right)\left(\alpha x_1 - \lambda\right)\right\}.$$

That is, with x_1 and x_3 held constant, V_2 must maximize the term in curved brackets. Differentation of this term with respect to V_2 yields the following

$$1 + \alpha V_2^* = \frac{x_2}{x_1}. \tag{6.40}$$

Comparison of (6.39) and (6.40) shows $V_2^* = V_1^*$. Continued application of (6.37) reveals all the V's to be equal. Hence we can write

$$f_N(x_1,\ x_{N+1}) = x_{N+1} - x_1 - N\lambda V^*$$

where V^* is the constant value of the V's, given by the following

$$(1 + \alpha V^*)^N = \frac{x_{N+1}}{x_1}. \tag{6.41}$$

Elimination of V^* and x_{N+1} in the expression for f_N by by means of equations (6.33) and (6.41) yields

$$f_N(x_1) = qx_f + rx_1 - x_1 - \frac{N\lambda}{\alpha}\left[\left(\frac{qx_f + rx_1}{x_1}\right)^{\frac{1}{N}} - 1\right].$$

Invoking (6.38), we maximize $f_N(x_1)$ with respect to x_1. This yields the relationship

$$(x_1^*)^2 = \frac{\lambda x_f}{\alpha}\left[r + \frac{qx_f}{x_1^*}\right]^{\frac{1-N}{N}}$$

or, in terms of V^*,

$$(1 + \alpha V^*)^{1-N}\left[(1 + \alpha V^*)^N - r\right]^2 = \frac{q^2\alpha x_f}{\lambda}. \tag{6.42}$$

For $r = 0$ (no recycle, $q = 1$), (6.42) yields the result obtained previously (6.23).

In the solution procedure used here, the first stage decision, V_1, was postponed (by considering x_1 fixed). Only when the effect of x_1 due to the recycle stream is incorporated into the functional equation can this decision be made.

Another way of looking at this procedure is to consider that the recycle loop is cut as depicted in Figure 6.8. This removes the cyclic character and effectively transforms the process into a serial one. The value of the *cut state*, x_1 in this case, becomes a decision variable, replacing V_1. The cut state also enters the process at stage N, since x_{N+1} depends on x_1. Hence the effect of x_1 on the objective function cannot be fully evaluated until stage N is reached.

At that point, the total effect of x_1 can be ascertained and the optimum decision on x_1 made.

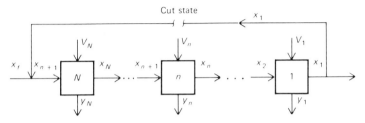

Figure 6.8

A similar procedure could be followed if the recycle loop contained stages where decisions had to be made. The cut state would be treated as a constant and the decisions made in such a way as to constitute an optimal policy with regard to the value assigned to the cut state. At the junction where the recycle loop returns to the main body of the process, the value of the recycle loop objective function would have to be incorporated with the value of the main body objective function in the functional equation. Then the decision as to the optimum cut state value can be made.

Loop structures play an important role in processes with recycle, feedback control systems, etc. A general treatment of the subject, together with many examples, is provided in the texts by Nemhauser and by Wilde and Beightler.

6.7 OTHER TECHNIQUES FOR STAGED SYSTEM

Although dynamic programming is admirably equipped to solve stagewise optimization problems, other techniques often furnish equally expedient solutions. For example, geometric programming can be used to solve the simple cross-current extraction problem. To show how the solution is obtained, we write the objective function as follows:

$$\text{Max } R = \sum \left\{ (x_{n+1} - x_n) - V_n \lambda \right\}$$

or $\qquad$ $$\text{Max } R = x_{N+1} - \left(x_1 + \lambda \sum_{n=1}^{N} V_n \right). \qquad (6.43)$$

Since x_{N+1} is the feed composition which is fixed, R is maximized in (6.43) if we minimize z, where

$$\text{Min } z = x_1 + \lambda \sum_{n=1}^{N} V_n \qquad\qquad (6.44)$$

$$(6.45)$$
subject to $x_{n+1} - x_n = V_n \alpha x_n \qquad n = 1, \ldots, N$

where the linear equilibrium has been assumed.

Equation (6.45) can be written as

$$\frac{x_n}{x_{n+1}} = \left(\frac{1}{V_n \alpha + 1} \right) \qquad n = 1, 2, \ldots, N.$$
$$(6.46)$$

Repetition of (6.46) N times gives

$$x_1 = x_{N+1} \left(\frac{1}{V_N \alpha + 1} \right) \cdots \left(\frac{1}{V_1 \alpha + 1} \right). \qquad (6.47)$$

Let $u_n = V_n \alpha + 1$. Then

$$x_1 = x_{N+1} \prod_{n=1}^{N} (u_n)^{-1}$$

and the objective function becomes

$$\text{Min } z = x_{N+1} \prod_{n=1}^{N} (u_n)^{-1} + \lambda \sum_{n=1}^{N} \left(\frac{u_n}{\alpha} \right) - N\lambda.$$
$$(6.48)$$

Although the last term has a negative sign, it is a constant and not subject to variation. The remaining portion of the objective function is a posynomial with $N+1$ terms and N unknowns. It forms a geometric programming problem of degree of difficulty zero. The dual function corresponding to (6.48) is

$$v(\bar{\delta}) = \left(\frac{x_{N+1}}{\delta_1} \right)^{\delta_1} \prod_{j=2}^{N+1} \left(\frac{\lambda}{\alpha\delta_j} \right)^{\delta_j}.$$

The orthogonality conditions are

$$- \delta_1 + \delta_j = 0 \qquad j = 2, \dots, N + 1$$

which indicates that the $\delta_j, j = 1, \dots, N + 1$ are equal. This condition implies that each of the corresponding terms in the objective function are equal

$$\frac{\lambda}{\alpha} u_1 = \frac{\lambda}{\alpha} u_2 = \dots = \frac{\lambda}{\alpha} u_N$$

and hence the V_n's are all equal.

The normality condition is

$$\delta_1 + \delta_2 + \dots + \delta_{N+1} = 1.$$

Hence
$$\delta_1 = \delta_2 = \dots = \delta_{N+1} = \frac{1}{N + 1}.$$

The dual function thus becomes

$$v(\bar{\delta}) = \frac{x_N^{\frac{1}{N+1}}}{(N + 1)} \left(\frac{\lambda}{\alpha} \right)^{\frac{N}{N+1}}$$

Since each term of the objective function contributes equally to $v(\bar{\delta})$, we find

$$\frac{\lambda}{\alpha} \ u_n^* = \frac{v(\bar{\delta})}{\delta_{n+1}} = x_N^{\frac{1}{N+1}} \left(\frac{\lambda}{\alpha} \right)^{\frac{N}{N+1}}$$

$$u_n^* = \left(\frac{\alpha x_N}{\lambda} \right)^{\frac{1}{N+1}} \tag{6.49}$$

which is, of course, the solution obtained by dynamic programming.

Geometric programming can be used to solve stagewise optimization problems if the nth stage index, r_n, is of form

$$r_n = x_{n+1} - x_n - \lambda_n \ V_n^{\alpha_n} \tag{6.50}$$

where α_n and λ_n are specified constants. If the transformation equations can be written as follows, with x_{N+1} given

$$x_n = T(V_n, \ x_{n+1}) = x_{n+1}(\mu_n \ V_n^{\beta_n}) \tag{6.51}$$

where β_n and μ_n are constants, then the objective function can be formulated as a posynomial:

$$\text{Min } z = x_{N+1} \prod_{n=1}^{N} \mu_n \ V_n^{\beta_n} + \sum_{n=1}^{N} \lambda_n \ V_n^{\alpha_n}.$$

This is now in the geometric programming format and is a problem with zero degree of difficulty.

Further generalizations of the stage returns and transformation equations can be made and still preserve the posynomial character of the modified performance index. For example, the returns and transformations could depend upon power functions of the past several decisions. This would present no undue difficulty for geometric programming, but it would require special techniques within the dynamic programming format.

Bibliography

Many excellent text and reference books on dynamic programming have appeared, although, as in linear programming, the total list of references is too long to include. The following books both summarize the method and report on most of the literature.

1. Aris, R. *The Optimal Design of Chemical Reactors.* Academic Press, New York, 1961.

2. Aris, R. *Discrete Dynamic Programming.* Blaisdell Publishing Co., New York, 1964.

3. Bellman, R. *Dynamic Programming.* Princeton University Press, Princeton, New Jersey, 1957.

4. Bellman, R. *Adaptive Control Processes: A Guided Tour,* Princeton University Press, Princeton, New Jersey, 1961.

5. Bellman, R. and S.E. Dreyfus. *Applied Dynamic Programming.* Princeton University Press, Princeton, New Jersey, 1962.

6. Nemhauser, George L. *Introduction to Dynamic Programming.* John Wiley & Sons, Inc., New York, 1967.

7. Roberts, S.M. *Dynamic Programming in Chemical Engineering and Process Control.* Academic Press, New York, 1964.

6.2 AN EXAMPLE OF A STAGEWISE PROCESS

This example was first presented in the literature by

8. Aris, R.B., D. Rudd and N.R. Amundson. *Chem. Eng. Sci.* 12, p. 88, 1960.

6.5 RELATIONSHIP OF DYNAMIC PROGRAMMING TO NONLINEAR PROGRAMMING

In addition to the treatment of this problem by Bellman, in the references cited above, Nemhauser has presented a clear development of the interrelationships between the two. George Hadley's presentation is also worthy of note:

9. Hadley, G. *Nonlinear and Dynamic Programming.* Addison-Wesley Publishing Co., Reading, Massachusetts, 1964.

6.6 EXTENSIONS TO NONSERIAL SYSTEMS

One of the first papers in this area was

10. Aris, R. G.L. Nemhauser and D.J. Wilde. Optimization of multistage cyclic and branching systems by serial procedures. *Amer. Inst. Chem. Eng., J.* 10, pp. 913—919, 1964.

The problem considered in the text appeared in the literature in the following:

11. Zahradnik, R.L. and D.H. Archer. A note on the application of the discrete maximum principle to cross-current extraction with a simple recycle stream. *I&EC Fund.* 3, pp. 232—234, 1964.

6.7 OTHER TECHNIQUES FOR STAGED SYSTEM

Among the alternatives to dynamic programming for solving staged systems is a technique known as the discrete maximum principle. This technique is better understood if the material in the next chapter is first read. Some references are

12. Butkovskii, A.G. The necessary and sufficient conditions for optimality of discrete control systems *Automat. i. Telemekh.* 24, pp. 1056—1064, 1963.

13. Fan, L.T. and C.S. Wang. *The Discrete Maximum Principle:* A Study of Multistage Systems Optimization. John Wiley and Sons, New York, 1964.

14. Halkin, H. Optimal Control for Systems Described by Difference Equations. *Advances in Control Systems: Theory and Applications.* (C.T. Leondes, ed.). Academic Press, New York, 1964.

15. Horn, F. and R. Jackson. Correspondence: Discrete maximum principle. *I&EC Fund.* 4, pp. 110—112, 1965.

16. Katz, S. A discrete version of Pontryagin's maximum principle. *J. Elec. Contr.* 5, pp. 179—184, 1962.

17. Katz, S. Best Operating Points for Staged Systems. *I&EC Fund.* 4, pp. 226—240, 1962.

18. Rozonoer, L.I., The Maximum Principle of L.S. Pontryagin in Optimal System Theory, Part III. *Automation and Remote Control*, 20, pp. 1519—1532, 1959.

19. Zahradnik, R.L. and D.H. Archer. Application of the discrete maximum principle to cross-current extraction. *I&EC Fund.* 2, pp. 238—240, 1963.

PART II
Trajectory
Optimization

7
Trajectory Optimization

7.1 INTRODUCTION

The focus of this chapter is on optimization problems in which the performance index is expressed as a function of a continuous variable. Such problems are common in many branches of engineering, such as mechanics, control theory, thermodynamics, etc. The chapter begins with several examples to illustrate the nature of these optimization problems and then develops the necessary conditions for their solutions. Techniques for obtaining the solutions which satisfy these conditions are discussed in chapter 8.

7.2 TRAJECTORY OPTIMIZATION PROBLEMS

Consider the general sequential allocation or stagewise optimization problem of the last chapter. Suppose the N stages are identified by the succeeding integer indices: t_1, t_2, . . . , t_N. The process is depicted on Figure 7.1. The decisions at each stage are V_{t_1}, V_{t_2}, . . . , V_{t_N} and the performance index of the system is

$$z = \sum_{n=1}^{N} r(x_{tn}, V_{tn}). \qquad (7.1)$$

The system transformation equations are

193

$$\frac{x_{tn+1} - x_{tn}}{t_{n+1} - t_n} = g(x_{tn}, V_{tn})$$

$$n = 0, 1, \ldots, N - 1 \qquad (7.2)$$

and

$$x_{t0} = x_0, \text{ given} \qquad (7.3)$$

where $t_{n+1} - t_n = 1$. The functions r and g are assumed to be twice continuously differentiable.

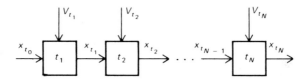

Figure 7.1

The decision process is depicted on Figure 7.2, part (a), where the N decisions are shown at the discrete points $t_1, t_2, \ldots, t_N$. Imagine now that the total number of stages remains fixed at N but that it is possible to make more than one decision per stage — a total of $M > N$ in all. Let the decisions be made at the points $\tau_1, \tau_2, \ldots, \tau_M$, as shown on part (b) of Figure 7.2. In this case, τ_m may be any value between t_0 and t_N.

In order to retain performance index (7.1), it will be necessary to normalize it as follows:

$$\frac{N}{M} \sum_{m=1}^{M} r(x_{\tau m}, V_{\tau m}). \qquad (7.4)$$

The interval between adjacent decision points is no longer unity, but rather

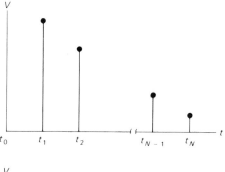

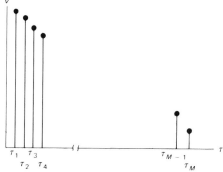

Figure 7.2

$$\tau_{m+1} - \tau_m = \frac{\tau_M - t_0}{M} = \frac{t_N - t_0}{M} = \frac{N}{M} \quad (7.5)$$

so that the system transformation equations become

$$\frac{x_{\tau m+1} - x_{\tau m}}{N/M} = g(x_{\tau m},\ V_{\tau m})$$

$$m = 1, 2, \ldots, M. \quad (7.6)$$

The initial condition (7.3) remains the same.

If now we pass to the limit of infinitely large M, the system index becomes an integral and the system equations become a single differential equation.

$$z = \int_{t_0}^{t_N} r[x(t), \ V(t)] \, dt \qquad (7.7)$$

$$\frac{dx}{dt} = g(x, \ V) \quad x(t_0) = x_0 \qquad (7.8)$$

The optimization problem is stated as follows: maximize the value of the definite integral (7.7) by selecting V as a function of the *continuous* variable t, $t_0 \leqslant t \leqslant t_N$, subject to the differential equation constraint (7.8). The variable t can represent many things — e.g., time, distance, etc.

Problems such as this are called *trajectory optimization problems*, since their solution requires that a complete decision trajectory be specified rather than values of the decision variable at certain specific locations. The mathematical discipline which deals with trajectory optimization problems is the calculus of variations.

Problems in which the decision variable must be obtained in terms of a continuous variable occur naturally in many engineering situations and are of more interest than merely the passage to a limit of parameter optimization problems.

Example 7.1 Optimal Temperature Profiles in a Tubular Reactor   Consecutive reactions of the following sort are often carried out in a plug flow tubular reactor.

$$A \underset{k_2}{\overset{k_1}{\rightleftarrows}} B \overset{k_3}{\to} C$$

If x_1 denotes the concentration of A and x_2 the concentration of B, the equations describing the rate of change of these compositions with respect to the normalized reactor length, t, $0 \leqslant t \leqslant 1$, are

$$\frac{dx_1}{dt} = -k_1 x_1 + k_2 x_2 \quad x_1(0) = x_{10}$$

$$\frac{dx_2}{dt} = k_1 x_1 - (k_2 + k_3)x_2 \quad x_2(0) = x_{20}$$

where k_1, k_2 and k_3 are temperature dependent rate constants. Typically

$$k_i = A_i e^{-E_i/T} \quad i = 1, 2, 3$$

where A_i and E_i are constants, and T = temperature.

If B is the desired product, a suitable objective function is the maximization of the outlet concentration of B by selecting $T(t)$, $0 \leq t \leq 1$

$$z = x_2(1) = \int_0^1 \frac{dx_2}{dt} \, dt$$

$$= \int_0^1 [k_1 x_1 - (k_2 + k_3)x_2] \, dt.$$

This problem is now in the format of a trajectory optimization problem.

Example 7.2 Optimal Control of a Dynamic System The dynamics of a control system can often be represented as a differential equation in terms of x, the deviation of the controlled variable from its desired set point, and V, the control action

$$\frac{dx}{dt} = g(x, V) \quad x(0) = x_0.$$

One would like to determine the control as a function of time, $t_0 \leq t \leq t_N$ in order to minimize the following *quadratic* performance index.

$$\int_{t_0}^{t_N} (x^2 + \rho \ V^2)dt$$

where $\rho > 0$ is a specified constant.

In many cases, the differential equation describing the system dynamics is linear in x and V, i.e.,

$$\frac{dx}{dt} = -ax + bV$$

where a and b are constants. In such cases, as we will see, the solution to the trajectory optimization problem can be expressed as a *feedback control law.*

Example 7.3 The Classic Brachistochrone Problem The classic problem in the calculus of variations was proposed in 1696 by John Bernoulli. The problem consists of specifying the shape of a frictionless wire strung between two points such that a bead of mass m sliding along the wire in a constant gravity field will proceed from one point to another in minimum time.

To illustrate the problem, consider two axes x and t drawn as in Figure 7.3 with increasing x pointing downward. If the origin $(0, 0)$ and $(x_1, \ t_1)$ represent the selected points and the velocity of the bead at the origin is zero, the following energy balance holds

$$1/2 \ m \ V^2 = mg \ x$$

Gain in Kinetic Energy   Loss in Potential Energy

where g = gravitational acceleration
and V = particle velocity.

If we let ϕ be the angle between the velocity and the horizontal, the velocity components are

$$\frac{dx}{d\theta} = V \ \cos \phi \ = \sqrt{2gx} \ \cos \phi$$

$$\frac{dt}{d\theta} = V \sin \phi = \sqrt{2gx} \sin \phi$$

where θ is time.
We wish to specify $\phi(\theta)$ $0 \leqslant \theta \leqslant \theta*$

$$\underset{\{\phi(\theta)\}}{\text{Min}} \int_0^{\theta*} d\theta = \underset{\{\phi(\theta)\}}{\text{Min}} \quad \theta*.$$

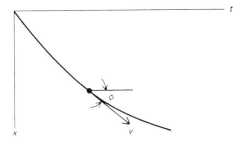

Figure 7.3

7.3 CALCULUS OF VARIATIONS

The calculus of variations had its beginning in the late seventeenth century with John Bernoulli's posing of the Brachistochrone problem. Since that time, the list of mathematicians who have contributed to its development reads like a Who's Who of mathematics: Newton, Leibnitz, Euler, Lagrange, Weierstrass, and Hilbert, to name only a few. Still, variational calculus is unable to meet the demands placed on it by its interested users. It continues today to be an active field of research.

One of the reasons for the continued development is that the calculus of variations has found many applications in engineering and science. These range from classical mechanics to modern-day control theory. In the next few sections, we will only be able to skim the surface of this fascinating discipline, but the reader is encouraged to con-

tinue his study by reference to the many texts which have
been written on the subject.

**7.3.1 The Simplest Problem of the Calculus of Varia-
tions** Consider again the trajectory optimization problem
posed in the last section

$$\text{Max } z = \int_{t_0}^{t_N} r(x, \ V) \ dt \qquad (7.9)$$

$$\frac{dx}{dt} = g(x, \ V) \quad x(t_0) = x_0 \qquad (7.10)$$

where f and g have continuous second derivatives in both
arguments.

As in the case of parameter optimization, we will first
establish the necessary conditions for an optimal solution.
In order to do this, consider that the optimal decision
policy is V^* and that this policy produces a state variable
trajectory x^*. A nonoptimal decision policy, V, may be
looked upon as a combination of V^* and an arbitrary but
smooth deviation $\xi(t)$; i.e.,

$$V = V^* + \xi(t). \qquad (7.11)$$

The state variable trajectory corresponding to V^* is x^*.
For any other control policy, we may express the state
variable path as

$$x = x^* + \epsilon\eta(t) \qquad (7.12)$$

where ϵ is an arbitrary constant and $\eta(t)$ is an arbitrary
function. However, since $x(t_0) = x_0$ for both an optimal
and nonoptimal decision policy, we must require that

$$\eta(t_0) = 0. \qquad (7.13)$$

ξ and η are related by the system differential equation,

$$\frac{dx}{dt} = \frac{d[x^* + \epsilon\eta]}{dt} = g[x^* + \epsilon\eta, \ V^* + \xi]. \quad (7.14)$$

Expansion of the right-hand side of (7.14) in the first terms of a Taylor series yields

$$\underline{\frac{dx^*}{dt}} + \epsilon \frac{d\eta}{dt} = \underline{g(x^*, \ V^*)}$$

$$+ \left(\frac{\partial g}{\partial x}\right)_* \epsilon\eta + \left(\frac{\partial g}{\partial V}\right)_* \xi \qquad (7.15)$$

where the subscript * on the partial derivatives means they are evaluated along the optimal trajectory.

The underlined parts of equation (7.15) are equal by virtue of the fact that the system differential equation must be satisfied by the optimal solution. Hence, (7.15) becomes

$$\epsilon \frac{d\eta}{dt} = \left(\frac{\partial g}{\partial x}\right)_* \epsilon\eta + \left(\frac{\partial g}{\partial V}\right)_* \xi. \quad (7.16)$$

If now $(\partial g/\partial V)_* \neq 0$, (7.16) may be used to relate ξ to η

$$\xi = \frac{\epsilon}{\left(\dfrac{\partial g}{\partial V}\right)_*} \left[\frac{d\eta}{dt} - \left(\frac{\partial g}{\partial x}\right)_* \eta\right]. \qquad (7.17)$$

We can use these same ideas to write the objective function, expanding the integrand about $(x^*, \ V^*)$:

$$\int_{t_0}^{t_N} r(x, \ V)dt =$$

$$\int_{t_0}^{t_N} \left(r(x^*, \ V^*) + \left(\frac{\partial r}{\partial x}\right)_* \epsilon\eta + \left(\frac{\partial r}{\partial V}\right)_* \xi\right) dt$$

or

$$\int_{t_0}^{t_N} r(x,\ V)dt\ -\ \int_{t_0}^{t_N} r(x^*,\ V^*)dt$$

$$=\ \int_{t_0}^{t_N} \left[\left(\frac{\partial r}{\partial x}\right)_* \epsilon\eta\ +\ \left(\frac{\partial r}{\partial V}\right)_* \xi \right] dt.$$

The left-hand side of this equation represents a nonmaximum value of the objective function minus the maximum value. Hence it is always less than or equal to zero. Thus

$$\int_{t_0}^{t_N} \left[\left(\frac{\partial r}{\partial x}\right)_* \epsilon\eta\ +\ \left(\frac{\partial r}{\partial V}\right)_* \xi \right] dt\ \leqslant\ 0.$$

Insertion of the expression for ξ into this inequality gives

$$\epsilon \int_{t_0}^{t_N} \left\{ \left[\left(\frac{\partial r}{\partial x}\right)_*\ -\ \left(\frac{\partial r}{\partial V}\right)_* \frac{\left(\frac{\partial g}{\partial x}\right)_*}{\left(\frac{\partial g}{\partial V}\right)_*} \right] \eta \right.$$

$$\left. +\ \frac{\left(\frac{\partial r}{\partial V}\right)_*}{\left(\frac{\partial g}{\partial V}\right)_*} \frac{d\eta}{dt} \right\} dt\ \leqslant\ 0.$$

Because ϵ is an arbitrary parameter which can assume either positive or negative values, the only way this inequality can be satisfied is for the integral to vanish.

$$\int_{t_0}^{t_N} \left\{ \left[\left(\frac{\partial r}{\partial x}\right)_* - \left(\frac{\partial r}{\partial V}\right)_* \frac{\left(\frac{\partial g}{\partial x}\right)_*}{\left(\frac{\partial g}{\partial V}\right)_*} \right] \eta \right.$$

$$\left. + \frac{\left(\frac{\partial r}{\partial V}\right)_*}{\left(\frac{\partial g}{\partial V}\right)_*} \frac{d\eta}{dt} \right\} dt = 0 \qquad (7.18)$$

In order to put this condition in more usable form, an integration by parts will be performed on the third term. Thus

$$\int_{t_0}^{t_N} \frac{\left(\frac{\partial r}{\partial V}\right)_*}{\left(\frac{\partial g}{\partial V}\right)_*} \frac{d\eta}{dt} dt = \left. \frac{\left(\frac{\partial r}{\partial V}\right)_*}{\left(\frac{\partial g}{\partial V}\right)_*} \eta \right|_{t_0}^{t_N}$$

$$- \int_{t_0}^{t_N} \eta \frac{d}{dt} \left[\frac{\left(\frac{\partial r}{\partial V}\right)_*}{\left(\frac{\partial g}{\partial V}\right)_*} \right] dt \qquad (7.19)$$

Notice that the integrated portion vanishes at the lower limit because of the fact that $\eta(t_0) = 0$ (7.13). In order for (7.18) to be satisfied, the upper limit must vanish as well; i.e.,

$$\left(\frac{\partial r}{\partial V}\right)_{*\ t\ =\ t_N} = 0. \qquad (7.20)$$

This condition is referred to as a *natural boundary condition* in variational calculus. It comes about if the value of

x at $t = t_N$ is not specified. If $x(t_N)$ were specified, $\eta(t_N)$ would have to be zero, and the integrated portion of the integral would vanish at the upper limit as well as the lower. In such a case, the natural boundary condition would not apply.

Combination of the integral in (7.19) with condition (7.18) gives

$$\int_{t_0}^{t_N} \left\{ \left(\frac{\partial r}{\partial x}\right)_* - \left(\frac{\partial r}{\partial V}\right)_* \frac{\left(\frac{\partial g}{\partial x}\right)_*}{\left(\frac{\partial g}{\partial V}\right)_*} \right. $$

$$\left. - \frac{d}{dt}\left[\frac{\left(\frac{\partial r}{\partial V}\right)_*}{\left(\frac{\partial g}{\partial V}\right)_*}\right] \right\} \eta\ dt = 0. \tag{7.21}$$

In order for this integral to vanish, it is necessary for the term in curved brackets to vanish for all t, $t_0 \leqslant t \leqslant t_N$. Because η is arbitrary, if the term in curved brackets did not vanish over some interval, η could be made positive in that interval and zero elsewhere. This would invalidate the equation. Hence, we require along an optimum trajectory that

$$\frac{d}{dt}\left[\frac{\left(\frac{\partial r}{\partial V}\right)}{\left(\frac{\partial g}{\partial V}\right)}\right] - \left[\frac{\partial r}{\partial x} - \left(\frac{\partial r}{\partial V}\right) \frac{\left(\frac{\partial g}{\partial x}\right)}{\left(\frac{\partial g}{\partial V}\right)}\right] = 0. \tag{7.22}$$

This is the first necessary condition for optimality. The necessary condition for an optimum in a parameter optimization problem, the vanishing of the first derivative, yields an algebraic equation whose solution provides the value of the parameters. In the case of trajectory optimization, the analogous condition yields a *differential equation* whose solution provides the desired trajectory. The differential equation is referred to as the Euler-Lagrange equation after two mathematicians who were involved in its development.

Although it may not be apparent, in order to obtain an optimal solution, the equivalent of a second-order differential equation must be solved. Two boundary conditions are required for the solution of second-order differential equations. In this case, they are

$$x(t_0) = x_0 \qquad (7.23)$$

and

$$\left(\frac{\partial r}{\partial V} \right)_{t = t_N} = 0. \qquad (7.20)$$

These conditions are specified at opposite ends of the given interval and thus constitute a two-point-boundary-value problem. The solution of such problems can be quite complicated, involving special mathematical or computational techniques.

Example 7.4 Obtain the optimum control policy $V(t)$, $t_0 \leqslant t \leqslant t_N$ which minimizes the following quadratic index

$$\frac{1}{2} \int_{t_0}^{t_N} (x^2 + V^2)dt$$

subject to

$$\frac{dx}{dt} = -ax + bV \qquad (i)$$

$$x(t_0) = x_0. \qquad (ii)$$

Solution   Here

$$r = \frac{1}{2} (x^2 + V^2); \frac{\partial r}{\partial x} = x; \frac{\partial r}{\partial V} = V,$$

$$g = -ax + bV; \frac{\partial g}{\partial x} = -a; \frac{\partial g}{\partial V} = b \neq 0$$

$$\frac{\dfrac{\partial r}{\partial V}}{\dfrac{\partial g}{\partial V}} = \frac{V}{b}; \frac{\dfrac{\partial g}{\partial x}}{\dfrac{\partial g}{\partial V}} = \frac{-a}{b}.$$

Substitution of these terms into equation (7.22) yields

$$\frac{d}{dt} \left(\frac{V}{b} \right) - \left[x - V \left(\frac{-a}{b} \right) \right] = 0$$

or

$$\frac{\partial V}{\partial t} = bx + aV. \tag{iii}$$

Boundary condition (7.20) implies that

$$V(t_N) = 0. \tag{iv}$$

Equations (i) and (iii) are two first-order differential equations — the equivalent of a second-order differential equations. Boundary conditions (ii) and (iv) at the two ends of the interval make the problem a two-point boundary value one.

This can be shown by differentiating (i) with respect to t:

$$\frac{d^2 x}{dt^2} = -a \frac{dx}{dt} + b \frac{dV}{dt}.$$

Substitution into this equation from (i) and (iii) yields

$$\frac{d^2x}{dt^2} = (a^2 + b^2)x. \qquad \text{(v)}$$

If we take $t_0 = 0$, the boundary conditions are

$$x(0) = x_0 \qquad \text{(vi)}$$

$$ax(t_N) + \left.\frac{dx}{dt}\right|_{t_N} = 0. \qquad \text{(vii)}$$

This second-order differential equation has the solution

$$x = x_0 \left[\frac{a \ \sinh \ m(t_N - t) + m \ \cosh \ m(t_N - t)}{a \ \sinh \ mt_N + m \ \cosh \ mt_N}\right]$$

where

$$m^2 = a^2 + b^2.$$

The classical expression of variational problems incorporates the differential equation into the objective function. This can be accomplished by writing (7.10) as follows

$$g(x, \ V) - \dot{x} = 0.$$

By the implicit function theorem, we know that this equation can be solved explicitly for V provided $\partial g/\partial V \neq 0$, a condition that has already been assumed. This relationship is expressed as

$$V = G(x, \ \dot{x}). \qquad \text{(7.24)}$$

(The dot symbolizes time differentiation; $\dot{x} = dx/dt$.) The implicit function theorem further provides (see Appendix B)

$$\frac{\partial G}{\partial \dot{x}} = \frac{1}{\dfrac{\partial g}{\partial V}} \quad \text{and} \quad \frac{\partial G}{\partial x} = -\frac{\dfrac{\partial g}{\partial x}}{\dfrac{\partial g}{\partial V}}. \qquad (7.25)$$

Substitution of the expression for V into the integrand gives

$$r(x, \ V) = r[x, \ G(x, \ \dot{x})] = R(x, \ \dot{x}).$$

The trajectory optimization problem is now stated as

$$\underset{\{x(t); \ t_0 \leqslant t \leqslant t_N\}}{\text{Max}} \quad \int_{t_0}^{t_N} R(x, \ \dot{x})dt. \quad (7.26)$$

By chain rule differentiation

$$\frac{\partial R}{\partial x} = \frac{\partial r}{\partial x} + \frac{\partial r}{\partial V} \frac{\partial G}{\partial x} \qquad (7.27)$$

$$\frac{\partial R}{\partial \dot{x}} = \frac{\partial r}{\partial V} \frac{\partial G}{\partial \dot{x}}. \qquad (7.28)$$

Elimination of the derivatives of G by means of (7.25) yields

$$\frac{\partial R}{\partial x} = \frac{\partial r}{\partial x} - \frac{\partial r}{\partial V} \frac{\left(\dfrac{\partial g}{\partial x}\right)}{\left(\dfrac{\partial g}{\partial V}\right)} \qquad (7.29)$$

$$\frac{\partial R}{\partial \dot{x}} = \frac{\partial r}{\partial V} \cdot \frac{1}{\left(\dfrac{\partial g}{\partial V}\right)}. \qquad (7.30)$$

Substitution of these last two equations into (7.22) results in the classical form of the Euler-Lagrange equation

$$\frac{d}{dt}\left[\frac{\partial R}{\partial \dot{x}}\right] - \frac{\partial R}{\partial x} = 0. \qquad (7.31)$$

And the natural boundary condition, if needed, is

$$\frac{\partial R}{\partial \dot{x}} = 0. \qquad (7.32)$$

Example 7.5 Re-express the Brachistochrone problem in a form of (7.26)

Solution By conservation of energy (example 7.3):

$$1/2 \; m \; V^2 = mgx. \qquad (i)$$

Also

$$V^2 = \left(\frac{dx}{d\theta}\right)^2 + \left(\frac{dt}{d\theta}\right)^2$$

$$V^2 = \left(\frac{dt}{d\theta}\right)^2 \left[1 + \left(\frac{dx}{dt}\right)^2\right]. \qquad (ii)$$

Combination of (i) and (ii) gives

$$\frac{dt}{d\theta} = \sqrt{\frac{2gx}{1 + (\dot{x})^2}}. \qquad (iii)$$

Since we wish to minimize the final time, we write with the aid of (iii):

$$\text{Min} \int_0^{\theta *} d\theta = \int_0^{t_1} \frac{d\theta}{dt} \; dt$$

$$= \int_0^{t_1} \sqrt{\frac{1 + (\dot{x})^2}{2gx}} \; dt. \qquad (iv)$$

The objective function is now in the form of equation (7.26). The integrand, R, is

$$R = \sqrt{\frac{1 + (\dot{x})^2}{2gx}}$$

and

$$\frac{\partial R}{\partial x} = -\frac{1}{2}\sqrt{\frac{1 + (\dot{x})^2}{2gx^3}}$$

$$\frac{\partial R}{\partial \dot{x}} = \frac{\dot{x}}{\sqrt{2gx(1 + (\dot{x})^2}} .$$

The Euler-Lagrange equation is

$$\frac{d}{dt}\left(\frac{\dot{x}}{\sqrt{2gx[1 + (\dot{x})^2]}}\right) + \frac{1}{2}\sqrt{\frac{1 + (\dot{x})^2}{2gx^3}} = 0 \quad \text{(v)}$$

with boundary conditions

$$\left.\begin{array}{l} x(0) = 0 \\[2mm] x(t_1) = x_1 \end{array}\right\} \text{specified by problem.}$$

In this case, the boundary conditions at both ends of the interval are specified, and there is no need to invoke the natural boundary condition.

The integration of the Euler-Lagrange equations is rarely a simple task, and more will be said about this later. In this case, equation (v) may be integrated to show that the optimal path is a cycloid, described by a point on a circle rolling on the x-axis.

If we expand the Euler-Lagrange equation (7.31), we obtain

$$\left(\frac{\partial^2 R}{\partial \dot{x} \partial \dot{x}}\right) \ddot{x} + \left(\frac{\partial^2 R}{\partial x \partial \dot{x}}\right) \dot{x} - \frac{\partial R}{\partial x} = 0. \qquad (7.33)$$

In order that this equation be a true second-order differential equation, it is necessary that the term multiplying $\ddot{x}$ not vanish along an extremal

$$\left(\frac{\partial^2 R}{\partial \dot{x} \partial \dot{x}}\right)_* \neq 0. \qquad (7.34)$$

Moreover, to insure that the solution of the Euler-Lagrange equation does provide a relative maximum for the functional, it is necessary that

$$\left(\frac{\partial^2 R}{\partial \dot{x} \partial \dot{x}}\right)_* \leqslant 0. \qquad (7.35)$$

This is known as the Legendre necessary condition. If the inequality holds for any value of x and $\dot{x}$, then this is also a sufficient condition for a relative maximum.

The variational problem posed by (7.26) can be generalized to allow for the possibility that the integrand may depend explicitly on the independent variable t; i.e.,

$$\begin{array}{c} \text{Max} \\ \{x(t),\ t_0 \leqslant t \leqslant t_N\} \end{array} \int_{t_0}^{t_N} R(x,\ \dot{x},\ t)dt. \ (7.36)$$

In this case, the Euler-Lagrange equation is formally the same as that given in (7.31). However, the expansion (7.33) now contains an extra term

$$\left(\frac{\partial^2 R}{\partial \dot{x} \partial \dot{x}}\right) \ddot{x} + \left(\frac{\partial^2 R}{\partial x \partial \dot{x}}\right) \dot{x} + \left(\frac{\partial^2 R}{\partial t \partial \dot{x}}\right) - \frac{\partial R}{\partial x} = 0. \quad (7.37)$$

7.3.2 Variational Problems with Integral Equality Constraints In many cases, the formulation of a trajectory optimization problem includes in addition to the integral to be maximized an equality constraint involving an integral. That is,

$$\text{Max} \quad \int_{t_0}^{t_N} R(x, \dot{x}, t)dt$$

subject to

$$\int_{t_0}^{t_N} P(x, \dot{x}, t)dt = \text{a constant.}$$

where the integrand function P, like R, is assumed to be twice continuously differentiable with respect to all of its arguments.

The solution of this problem can be obtained with the aid of Lagrange multipliers. The technique consists of multiplying the integrand of the constraint integral by an unknown constant λ and adding the product to the objective function integrand; thus

$$\int_{t_0}^{t_N} (R + \lambda P)dt.$$

This problem is then treated as an unconstrained one. The Euler-Lagrange equation

$$\frac{d}{dt}\left[\frac{\partial}{\partial \dot{x}}(R + \lambda P)\right] - \frac{\partial}{\partial x}(R + \lambda P) = 0$$

yields a solution in terms of λ; λ is chosen to satisfy the constraint.

Example 7.6 (Isoperimetric Problem) Find the closed curve of given circumference which maximizes the enclosed area.

Solution If we assume the curve to be divided into two equal parts by the t-axis, then we have

$$\text{Max Area} = \int_0^{t_N} x\,dt$$

subject to

$$\text{Perimeter} = \int_0^{t_N} \sqrt{1 + (\dot{x})^2}\;dt = \ell, \text{ the length.}$$

Here,

$$R = x \text{ and } P = \sqrt{1 + (\dot{x})^2}.$$

Hence, the Euler-Lagrange equation is

$$\frac{d}{dt}\left[\frac{\lambda\,\dot{x}}{\sqrt{1 + (\dot{x})^2}}\right] - 1 = 0,$$

from which we obtain by integration

$$\frac{\lambda\,\dot{x}}{\sqrt{1 + (\dot{x})^2}} = t + c \tag{i}$$

where c is a constant of integration.

Rearrangement of (i) gives

$$\dot{x} = \frac{dx}{dt} = \frac{t + c}{\sqrt{\lambda^2 - (t + c)^2}}. \tag{ii}$$

The integration of (ii) reveals the extremals to be circles. λ is then selected to provide the circle with the stated circumference.

In addition to integral equality constraints, trajectory optimization problems may have differential equation constraints or inequality constraints on the state or decision variables. These problems can be treated by classical variational calculus, but they are handled more conveniently in the modern context of the maximum principle and will be discussed in that section.

7.3.3 Other Generalizations

There are several generalizations of the simplest problem of variational calculus. One such generalization is the occurrence of more than one state variable. Suppose there are n, x_1, x_2, . . . , x_n. Then the objective function would have the form

$$\int_{t_0}^{t_N} R(\bar{x}, \dot{\bar{x}}, t)dt$$

and there would be an Euler-Lagrange equation for each variable:

$$\frac{d}{dt}\left(\frac{\partial R}{\partial \dot{x}_i}\right) - \frac{\partial R}{\partial x_i} = 0 \quad i = 1, 2, \ldots, n.$$

Suitable boundary conditions would have to be supplied or natural boundary conditions invoked for all x_i, $i = 1$, 2, . . . , n. Problems of this kind occur frequently in modern day systems applications.

A second generalization occurs when there are more than one independent variable in the problem statement. In the two-dimensional case, the dependent variable is a function of the variables s and t. The objective function is

$$\int_{s_0}^{s_N} \int_{t_0}^{t_N} R(x, \frac{\partial x}{\partial s}, \frac{\partial x}{\partial t}, s, t)dtds.$$

The Euler-Lagrange equation for this problem can be derived in a manner similar to that for a single independent variable. In this case, however, it is a partial differential equation

$$\frac{\partial}{\partial s}\left[\frac{\partial R}{\partial\left(\frac{\partial x}{\partial s}\right)}\right] + \frac{\partial}{\partial t}\left[\frac{\partial R}{\partial\left(\frac{\partial x}{\partial t}\right)}\right] - \frac{\partial R}{\partial x} = 0. \quad (7.38)$$

Again, suitable boundary conditions must be specified or obtained as natural conditions.

Variational problems in more than one independent variable occur in many areas of mechanics. They find modern day expression in the Schrödinger equation.

Example 7.7 Find the Euler-Lagrange equation which minimizes the following integral

$$\frac{1}{2}\int_0^{s_N}\int_0^{t_N}\left[\left(\frac{\partial u}{\partial s}\right)^2 + \left(\frac{\partial u}{\partial t}\right)^2\right]dsdt.$$

Solution Application of (7.38) to the problem yields

$$\frac{\partial^2 u}{\partial s^2} + \frac{\partial^2 u}{\partial t^2} = 0.$$

This is the Laplace or potential equation which describes, for example, the steady-state conduction of heat in a slab. The fact that this equation provides the minimum for the stated integral is an interesting consequence of variational mathematics and is known as Dirichlet's principle. Similar variational principles have been posed for many of the equations of physics and chemistry. For further discussion on this topic, the reader is referred to texts (9), (10) and (11).

7.4 *CONTINUOUS DYNAMIC PROGRAMMING*

Many of the equations and conditions of classical calculus of variations can be developed within the framework of dynamic programming. This approach leads to a slightly different interpretation of the principal results of variational calculus and hence is a useful exercise. Moreover, in some applications dynamic programming may provide an alternative computational approach of greater efficiency than straightforward solution of the Euler-Lagrange equations.

The equations characterizing an extremal can be obtained by application of the principle of optimality and utilization of the properties of continuous functions. Consider the trajectory optimization problem posed in Section 7.2:

$$\underset{\{V(t)\}}{\text{Max}} \int_{t_0}^{t_N} r(x, V)dt \tag{7.39}$$

subject to

$$\frac{dx}{dt} = g(x, V) \quad x(t_0) = x_0. \tag{7.40}$$

If the upper limit on the objective function integral is considered fixed, the maximum value of the objective function depends only upon the lower limit t_0 and the value of x at this limit, x_0. In the same manner as discrete dynamic programming, we call this value $f(x_0, t_0)$

$$f(x_0, t_0) = \underset{\{V(t)\}}{\text{Max}} \int_{t_0}^{t_N} r(x, V)dt. \tag{7.41}$$

The integral may be written as the sum of two integrals, the first between t_0 and $t_0 + \Delta t$, the second between $t_0 + \Delta t$ and t_N, with Δt understood to be small.

$$\text{Max} \int_{t_0}^{t_N} r(x, V)dt = \text{Max} \left[\underbrace{\int_{t_0}^{t_0 + \Delta t} r(x, V)dt}_{1} \right.$$

$$\left. + \underbrace{\int_{t_0 + \Delta t}^{t_N} r(x, V)dt}_{2} \right]$$

The mean value theorem for integrals assures that the first part of this expansion can be written as $\bar{r}(x, V)\Delta t$ where the bar signifies that $r(x, V)$ is to be evaluated at some x and V value between t_0 and $t_0 + \Delta t$.

The second term of the expansion is identified as

$$f(x_0 + \Delta x, t_0 + \Delta t) = \int_{t_0 + \Delta t}^{t_N} r(x, V)dt$$

where $x_0 + \Delta x$ is the value of x at $t_0 + \Delta t$.

Substitution of these terms into (7.41) yields

$$f(x_0, t_0) = \underset{\substack{\{V(t)\} \\ t_0 \leq t \leq t_0 + \Delta t}}{\text{Max}} [\bar{r}(x, V)\Delta t + f(x_0 + \Delta x, t_0 + \Delta t)]. \tag{7.42}$$

Note the similarity of (7.42) with its discrete analog, (6.18).

In the continuous case, it is possible to utilize the continuity of f to expand the second term in brackets in (7.42) in a Taylor series about the point (x_0, t_0). This gives

$$f(x_0, t_0) = \underset{\substack{\{V(t)\} \\ t_0 \leq t \leq t_0 + \Delta t}}{\text{Max}} \left[\bar{r}(x, V)\Delta t + f(x_0, t_0) \right.$$

$$+ \left(\frac{\partial f}{\partial x}\right)_0 \Delta x + \left(\frac{\partial f}{\partial t}\right)_0 \Delta t + 0(\Delta x^2) + 0(\Delta t^2)\Bigg]$$

where the subscripts o on the partial derivatives indicate they are to be evaluated at the point $(x_0,\ t_0)$.

Since $f(x_0,\ t_0)$ inside the brackets is not subject to the maximization procedure, it cancels the similar term on the left, to yield

$$0 = \underset{\substack{\{V\ (t)\} \\ t_0 \leqslant t \leqslant t_0 + \Delta t}}{\text{Max}} \Bigg[\bar{r}(x,\ V)\Delta t + \left(\frac{\partial f}{\partial x}\right)_0 \Delta x + \left(\frac{\partial f}{\partial t}\right)_0 \Delta t$$

$$+ 0(\Delta x^2) + 0(\Delta t^2)\Bigg]. \qquad (7.43)$$

The term Δx is related to Δt by (7.40)

$$\Delta x = \bar{g}(x,\ V)\Delta t \qquad (7.44)$$

where the bar over the function g carries the same interpretation as that over r. Substitution of (7.44) into (7.43) yields

$$0 = \underset{\substack{\{V\ (t)\} \\ t_0 \leqslant t \leqslant t_0 + \Delta t}}{\text{Max}} \Bigg[\bar{r}(x,\ V)\Delta t + \left(\frac{\partial f}{\partial x}\right)_0 \bar{g}(x,\ V)\Delta t$$

$$+ \left(\frac{\partial f}{\partial t}\right)_0 \Delta t + 0(\Delta t^2)\Bigg]. \qquad (7.45)$$

This equation is divided by Δt and then Δt passed to the limit zero. This results in

$$0 = \text{Max}_{\{V (t_0)\}} \left[r[x_0,\ V(t_0)] + \left(\frac{\partial f}{\partial x}\right)_0 g[x_0,\ V(t_0)] \right.$$

$$\left. + \left(\frac{\partial f}{\partial t}\right)_0 \right]. \tag{7.46}$$

There is an obvious corollary of the principle of optimality which states that the decision on any portion of an optimal trajectory must also be optimal. Hence (7.46) holds not just for t_0 but for all t, $t_0 \leqslant t \leqslant t_N$:

$$0 = \text{Max}_{\substack{\{V (t)\} \\ t_0 \leqslant t \leqslant t_N}} \left[r[x,\ V(t)] + \left(\frac{\partial f}{\partial x}\right) g[x,\ V(t)] \right.$$

$$\left. + \frac{\partial f}{\partial t} \right]. \tag{7.47}$$

Also $\partial f/\partial t$ is not involved explicitly in the maximization process, so we have finally

$$-\frac{\partial f}{\partial t} = \text{Max}_{\{V (t)\}} \left[r[x,\ V(t)] + \frac{\partial f}{\partial x} g[x,\ V(t)] \right]. \tag{7.48}$$

Equation (7.48) is the basic functional equation of continuous dynamic programming. It is a nonlinear first-order partial differential equation. Its solution is a *surface* in which the optimal trajectory is imbedded — analogous to the discrete case.

There are several methods for solving (7.48). One consists in replacing the continuous variables and partial derivatives by finite difference approximations. Alternatively, the method of characteristics could be used to solve the equation. In most cases, one can simply carry out the maximization operation suggested by (7.48) and, with proper manipulation, arrive at an equation in terms of x and t.

Example 7.8 Restate the Brachistochrone problem and develop its solution by means of the functional equation of continuous dynamic programming.

Solution The problem is restated as follows

$$\text{Min} \quad \int_0^{t_1} \sqrt{\frac{1 + V^2}{2 \, gx}} \; dt. \qquad \text{(i)}$$

subject to

$$\frac{dx}{dt} = V. \qquad \text{(ii)}$$

Hence, (7.48) becomes

$$-\frac{\partial f}{\partial t} = \underset{V \, (t)}{\text{Min}} \left[\sqrt{\frac{1 + V^2}{2 \, gx}} + \frac{\partial f}{\partial x} \, V \right].$$

For the bracketed term to be minimized, it is necessary that

$$\frac{V}{\sqrt{2 \, gx(1 + V^2)}} + \frac{\partial f}{\partial x} = 0$$

or

$$\left(\frac{\partial f}{\partial x} \right) = - \frac{V}{\sqrt{2 \, gx(1 + V^2)}} \qquad \text{(iii)}$$

and

$$\left(\frac{\partial f}{\partial t}\right) = \frac{-1}{\sqrt{2\ gx(1\ +\ V^2)}} \ .\qquad\text{(iv)}$$

In order to eliminate f from these equations, we note that

$$\frac{d}{dt}\left(\frac{\partial f}{\partial x}\right) = \frac{\partial^2 f}{\partial t\partial x} + \frac{\partial^2 f}{\partial x^2}\frac{dx}{dt}$$

or

$$\frac{\partial^2 f}{\partial t\partial x} = \frac{d}{dt}\left(\frac{\partial f}{\partial x}\right) - \frac{\partial^2 f}{\partial x^2}\frac{dx}{dt}.$$

Hence, from equation (iii),

$$\frac{\partial^2 f}{\partial t\partial x} = \frac{-d}{dt}\left(\frac{V}{\sqrt{2\ gx(1\ +\ V^2)}}\right)$$

$$-\ \frac{V\dot{x}}{2\sqrt{2\ gx^3\ (1\ +\ V^2)}}.\qquad\text{(v)}$$

Partial differentiation of equation (iv) with respect to x yields

$$\frac{\partial^2 f}{\partial x\partial t} = \frac{1}{2\sqrt{2\ gx^3\ (1\ +\ V^2)}}.\qquad\text{(vi)}$$

222   TRAJECTORY OPTIMIZATION

By equating equations (v) and (vi) and using equation (ii), we obtain

$$\frac{d}{dt}\left(\frac{\dot{x}}{\sqrt{2\,gx(1+\dot{x}^2)}}\right)+\frac{1}{2}\sqrt{\frac{1+\dot{x}^2}{2\,gx^3}}=0$$

which is the same equation as we obtained earlier.

The solution procedure from this example can be formalized by noting that (7.48) is equivalent to the two simultaneous equations

$$-\frac{\partial f}{\partial t}=r(x,\ V)+\frac{\partial f}{\partial x}\,g(x,\ V)\qquad(7.49)$$

and

$$\frac{\partial r}{\partial V}+\frac{\partial f}{\partial x}\,\frac{\partial g}{\partial V}=0\qquad(7.50)[$$

where the last equation expresses the necessary condition for the maximization required in (7.48).

These equations can be rewritten as

$$-\frac{\partial f}{\partial t}=r(x,\ V)-\frac{\left(\dfrac{\partial r}{\partial V}\right)}{\left(\dfrac{\partial g}{\partial V}\right)}g(x,\ V)\qquad(7.51)$$

$$-\frac{\partial f}{\partial x}=-\frac{\left(\dfrac{\partial r}{\partial V}\right)}{\left(\dfrac{\partial g}{\partial V}\right)}\ .\qquad(7.52)$$

Partial differentiation of (7.51) with respect to x gives

$$- \frac{\partial^2 f}{\partial x \partial t} = \frac{\partial r}{\partial x} - \frac{\partial g}{\partial x} \left(\frac{\frac{\partial r}{\partial V}}{\frac{\partial g}{\partial V}} \right) - g(x, V) \frac{\partial}{\partial x} \left(\frac{\frac{\partial r}{\partial V}}{\frac{\partial g}{\partial V}} \right). \quad (7.53)$$

Total differentiation of (7.52) with respect to t gives

$$- \frac{d}{dt} \left(\frac{\partial f}{\partial x} \right) = - \frac{\partial^2 f}{\partial t \partial x} - \frac{\partial}{\partial x} \left(\frac{\partial f}{\partial x} \right) \dot{x}$$

$$= - \frac{d}{dt} \left(\frac{\left(\frac{\partial r}{\partial V} \right)}{\left(\frac{\partial g}{\partial V} \right)} \right). \quad (7.54)$$

Rearrangement of (7.54) provides, with the aid of (7.52) and (7.40).

$$- \frac{\partial^2 f}{\partial t \partial x} = - \frac{\partial}{\partial x} \left(\frac{\left(\frac{\partial r}{\partial V} \right)}{\left(\frac{\partial g}{\partial V} \right)} \right) g(x, V) - \frac{d}{dt} \left(\frac{\left(\frac{\partial r}{\partial V} \right)}{\left(\frac{\partial g}{\partial V} \right)} \right).$$

$$(7.55)$$

Subtraction of (7.55) from (7.53) gives

$$\frac{d}{dt}\left(\frac{\left(\frac{\partial r}{\partial V}\right)}{\left(\frac{\partial g}{\partial V}\right)}\right) - \left[\frac{\partial r}{\partial x} - \left(\frac{\partial r}{\partial V}\right)\frac{\left(\frac{\partial g}{\partial x}\right)}{\left(\frac{\partial g}{\partial V}\right)}\right] \qquad (7.56)$$

which is the Euler-Lagrange equation.

Since continuous dynamic programming leads to the Euler-Lagrange equation as a solution procedure, in most cases there is no need to go through the formalism of the basic functional equation. If, however, there are inequality constraints on either the state or decision variables, then the Euler-Lagrange equation may not properly identify the extremal and one is forced to the original development.

Alternatively, there is another way of stating the necessary conditions for a trajectory optimization problem. It is known as the maximum principle.

7.5 THE MAXIMUM PRINCIPLE

The basic functional equation of continuous dynamic programming is

$$\frac{\partial f}{\partial t} = \underset{\{V(t)\}}{\text{Max}}\left[r(x, V) + \frac{\partial f}{\partial x} g(x, V)\right]. \qquad (7.57)$$

Instead of retaining the partial derivative of f with respect to x, we identify it as $\lambda(t)$.

$$\lambda(t) = \frac{\partial f}{\partial x} \qquad (7.58)$$

Also, the term in brackets to be maximized is symbolized by $H(x, V, \lambda)$:

$$H(x, V, \lambda) = r(x, V) + \lambda(t)g(x, V). \qquad (7.59)$$

If the derivative of H with respect to V is set equal to zero, we have shown that

$$\frac{\partial f}{\partial x} = - \frac{\dfrac{\partial r}{\partial V}}{\dfrac{\partial g}{\partial V}} . \qquad (7.60)$$

Hence, the Euler-Lagrange equation may be written as

$$\frac{d\lambda}{dt} = - \frac{\partial r}{\partial x} - \lambda \frac{\partial g}{\partial x} .$$

Boundary condition (7.20) implies that

$$\lambda(t_N) = 0.$$

Thus, we have an alternative way of stating the necessary condition for a trajectory optimization problem. To summarize,

$$\text{Max} \quad \int_{t_0}^{t_N} r(x, V)dt$$

subject to

$$\dot{x} = g(x, V) \quad x(t_0) = x_0 \qquad \begin{array}{l} \text{Problem} \\ \text{Statement} \end{array}$$

$$\underset{\{V(t)\}}{\text{Max}} \quad H = r(x, V) + \lambda g(x, V)$$

where

$$\frac{d\lambda}{dt} = - \frac{\partial r}{\partial x} - \lambda \frac{\partial g}{\partial x} \qquad \begin{array}{l} \text{Necessary} \\ \text{Condition} \end{array}$$

and

$$\lambda(t_N) = 0. \qquad (7.61)$$

The function to be maximized, H, is called a *Hamiltonian* and the variable λ is called an *adjoint variable*. The statement of the necessary conditions in these terms is called the *maximum principle.*

The maximum principle, in a sense, resembles the Lagrangian approach to parameter optimization problems, and indeed the adjoint variables can be interpreted as Lagrange multipliers. The equations of continuous dynamic programming contain explicit formulae for these multipliers, much like the differential approach in parameter optimization.

The maximum principle was developed by a Russian mathematician, L.S. Pontryagin, in the 1950s. Pontryagin's development did not follow from the functional equation of dynamic programming in the manner shown here but rather was based on variational arguments such as were used earlier.

The derivation of the necessary conditions stated in (7.61) utilized the vanishing of the first derivative of H as a necessary condition for maximization. From our experience with parameter optimization, we know that such a condition does not guarantee that H is indeed maximized, as called for by the functional equation of continuous dynamic programming. The only firm conclusion possible is that the necessary conditions (7.61) must be interpreted as requiring that the derivative of H with respect to V must vanish. In such a form, they are referred to as the *weak maximum principle.*

In many cases of practical interest, it is desirable to have a statement of the necessary conditions for trajectory optimization which identify the Hamiltonian as maximum or minimum. This is particularly true when inequality constraints are placed on the decision variables — e.g.,

$$0 \leqslant V \leqslant 1. \qquad (7.62)$$

To derive stronger necessary conditions for this problem, consider again that V^* represents the optimal decision policy and x^* the corresponding optimal state trajectory. We consider a perturbation in the decision policy, ξ, which occurs at $t = \bar{t}$ and lasts for a duration Δt. Hence the comparison decision policy, V, is

$$V = \begin{cases} V^* + \xi, & \bar{t} \leqslant t \leqslant \bar{t} + \Delta t \\ V^* & \text{elsewhere.} \end{cases}$$

This perturbation produces a change in the state variable which continues for the duration of the process; i.e.,

$$x = \begin{cases} x^* & , & t_0 \leqslant t \leqslant \bar{t} \\ x^* + \eta, & \bar{t} \leqslant t < t_N. \end{cases}$$

The previous change in V was infinitesimal but persisted over the entire range of t. The present change is finite in magnitude but lasts only over a small interval.

The change in the objective function produced by the perturbation in V is

$$z^* - z = \int_{\bar{t}}^{\bar{t} + \Delta t} [r(x^*, \ V^*) - r(x, \ V)]\, dt$$

$$+ \int_{\bar{t} + \Delta t}^{t_N} [r(x^*, \ V^*) - r(x, \ V)]\, dt \qquad (7.63)$$

The perturbation η is related to the perturbation ξ by the system equation

$$\frac{d\eta}{dt} = g(x, \ V) - g(x^*, \ V^*).$$

In this case, it is desirable to expand the function about the point $g(x^*, V)$. This gives the following first-order expansion

$$\frac{d\eta}{dt} = g(x^*, V) + \left(\frac{\partial g}{\partial x}\right)\eta - g(x^*, V^*), \quad \bar{t} \leqslant t \leqslant t_N$$

(7.64)

where the partial derivative is not necessarily evaluated along the extremal. Equation (7.64) is multiplied by λ and rearranged

$$\lambda\left(\frac{\partial g}{\partial x}\right)\eta = \lambda[g(x^*, V^*) - g(x^*, V)] + \lambda\frac{d\eta}{dt}.$$

(7.65)

The adjoint variable λ has the same meaning here as it did in the previous case. Whether V^* is the unconstrained maximum of H or V^* is at one of its limits, the derivative of H with respect to V still vanishes along an extremal path. Hence, (7.61) for λ is still valid. Multiplication of this equation by η gives

$$-\frac{\partial r}{\partial x}\eta = \eta\frac{d\lambda}{dt} + \lambda\frac{\partial g}{\partial x}\eta.$$

(7.66)

Substitution for the last term from (7.65) gives

$$-\frac{\partial r}{\partial x}\eta = \eta\frac{d\lambda}{dt} + \lambda[g(x^*, V^*) - g(x^*, V)] + \lambda\frac{d\eta}{dt}$$

(7.67)

This equation holds for the interval $\bar{t} \leqslant t \leqslant t_N$. However, for $\bar{t} + \Delta t \leqslant t \leqslant t_N$, $V = V^*$ and the square-bracketed term vanishes.

Returning now to the change in the objective function, we expand $r(x, V)$ about (x^*, V) to yield

$$z^* - z = \int_{\bar{t}}^{\bar{t} + \Delta t} \left(r(x^*, V^*) - r(x^* \; V) - \frac{\partial r}{\partial x} \; \eta \right)$$

$$+ \int_{\bar{t} + \Delta t}^{t_N} \left(r(x^*, V^*) - r(x^*, V) - \frac{\partial r}{\partial x} \; \eta \right) \, dt.$$

Substitution of the expression for $- \partial r / \partial x \; \eta$ into the two integral parts of this equation gives the following. Again note that $V = V^*$ in the interval $\bar{t} + \Delta t \leqslant t \leqslant t_N$, so the two r-values in the second integral are equal

$$z^* - z = \int_{\bar{t}}^{\bar{t} + \Delta t} \Big\{ r(x^*, V^*) - r(x^*, V)$$

$$+ \; \lambda \; [g(x^*, V^*) - g(x^*, V)] + \frac{d}{dt} \; (\lambda \eta) \Big\} \; dt$$

$$+ \int_{\bar{t} + \Delta t}^{t_N} \frac{d}{dt} \; (\eta \lambda) dt. \tag{7.68}$$

where we have utilized the fact that

$$\eta \; \frac{d\lambda}{dt} + \lambda \; \frac{d\eta}{dt} = \frac{d}{dt} \; (\eta \lambda).$$

Rearrangement of (7.68) gives

$$z^* - z = \int_{\bar{t}}^{\bar{t} + \Delta t} \Big\{ r(x^*, V^*) + \lambda \; g(x^*, V^*)$$

$$- [r(x^*, \ V) + \lambda \ g(x^*, \ V)] \Big\} \ dt$$

$$+ \int_{\bar{t}}^{t_N} \frac{d}{dt} \ (\eta\lambda)dt.$$

The last integral, which may be integrated directly, vanishes since $\eta(\bar{t}) = 0$ and $\lambda(t_N) = 0$ (7.61). Since $z^* - z \geqslant 0$, the remaining integral must also be non-negative. Passing then to the limit as $\Delta t \to 0$, we conclude that at $\bar{t}$

$$r(x^*, \ V^*) + \lambda \ g(x^*, \ V^*) \geqslant r(x^*, \ V) + \lambda \ g(x^*, \ V).$$

In other words, the Hamiltonian evaluated along the extremal is greater than the Hamiltonian evaluated elsewhere. Hence, the optimal decision policy does maximize H. This is a statement of the *strong* maximum principle. If the objective function is to be minimized, then the Hamiltonian must be correspondingly minimized.

A number of proofs of the strong maximum principle have been published, which involve varying degrees of mathematical sophistication (*14, 15*).

7.51 Extensions to Large Systems   One advantage of the maximum principle is that it extends easily to larger systems. Consider a trajectory optimization problem involving a single decision variable, V, but n state variables, $x_1, \ x_2, \ . \ . \ . \ , \ x_n$. The optimization problem is now stated as follows. Select $V(t), \ t_0 \leqslant t \leqslant t_N$ to maximize the following integral:

$$\int_{t_0}^{t_N} r(\bar{x}, \ V)dt$$

while satisfying

$$\frac{dx_i}{dt} = g_i(\bar{x}, \ V) \quad x_i(t_0) = x_{i0} \quad i = 1, \ 2, \ . \ . \ . \ , \ n.$$

The maximum principle states that the optimal $V(t)$ must maximize for all t, $t_0 \leqslant t \leqslant t_N$ the following Hamiltonian expression

$$H = r(\bar{x}, V) + \sum_{j=1}^{n} \lambda_j g_j(\bar{x}, V)$$

where

$$\frac{d\lambda_i}{dt} = -\frac{\partial r}{\partial x_i} - \sum_{j=1}^{n} \lambda_j \frac{\partial g_j}{\partial x_i}$$

and

$$\lambda_i(t_N) = 0 \quad i = 1, 2, \ldots, n$$

provided no conditions on $x_i(t_N)$ are specified.

Example 7.9 Minimum Time Problem A problem of considerable interest in modern control applications is the so-called minimum time problem. Simply stated in terms of two variables, the problem consists of specifying a control policy, $V(t)$, which will return the following dynamic system to the origin in minimum time:

$$\frac{dx_i}{dt} = a_{11} x_1 + a_{12} x_2 + b_1 V; \qquad x_1(0) = x_{10}$$

$$\frac{dx_2}{dt} = a_{21} x_1 + a_{22} x_2 + b_2 V; \qquad x_2(0) = x_{20}$$

subject to

$$-1 \leqslant V \leqslant +1.$$

Solution The integrand of the objective function is 1, so the Hamiltonian is

$$H = 1 + \lambda_1 (a_{11} x_1 + a_{12} x_2 + b_1 \ V)$$

$$+ \lambda_2 (a_{21} x_1 + a_{22} x_2 + b_2 \ V)$$

where

$$\frac{d\lambda_1}{dt} = -a_{11}\lambda_1 - a_{21}\lambda_2$$

$$\frac{d\lambda_2}{dt} = -a_{12}\lambda_1 - a_{22}\lambda_2.$$

Since $\bar{x}$ values are stated for both ends of the interval, no boundary conditions are specified on $\bar{\lambda}$.

Notice that the Hamiltonian is linear in V. Hence in order to minimize H, we must examine the sign of V.

$$V = +1 \text{ if } \lambda_1 b_1 + \lambda_2 b_2 < 0$$

$$V = -1 \text{ if } \lambda_1 b_1 + \lambda_2 b_2 > 0$$

If $\lambda_1 b_1 + \lambda_2 b_2 = 0$, no decision can be made. Such problems are referred to as *singular* problems. In this case, it can be shown that $\lambda_1 b_1 + \lambda_2 b_2$ can be zero for no more than an instant. The optimal control policy is thus to have V at either one or other of its limits. Such control is referred to as bang-bang control.

In the case we have been considering, namely one in which t does not appear directly, the Hamiltonian is a constant. This is easily shown by differentiating H with respect to the independent variable t:

$$\frac{dH}{dt} = \frac{dr}{dt} + \lambda \ \frac{dg}{dt} + g \ \frac{d\lambda}{\partial t}.$$

By chain rule differentiation,

$$\frac{dH}{dt} = \frac{\partial r}{\partial x} \ \dot{x} + \frac{\partial r}{\partial V} \ \dot{V} + \lambda \ \frac{\partial g}{\partial x} \ \dot{x} + \lambda \ \frac{\partial g}{\partial V} \ \dot{V} + g \ \frac{d\lambda}{dt}$$

after rearranging,

$$\frac{dH}{dt} = \left[\frac{\partial r}{\partial x} + \lambda \frac{\partial g}{\partial x} + \frac{d\lambda}{dt} \right] \dot{x} + \left[\frac{\partial r}{\partial V} + \lambda \frac{\partial g}{\partial V} \right] \dot{V}.$$

The first bracketed term is the adjoint equation, and hence it vanishes. The second bracketed term is the derivative of the Hamiltonian with respect to the decision variable, and it also vanishes. Hence,

$$\frac{dH}{dt} = 0; \qquad H = \text{constant}.$$

For trajectory problems in which the final value of the independent variable is not specified, this constant is zero.

Often, the objective function is expressed in the following implicit way

$$\text{Max} \sum_{i=1}^{n} c_i x_i(t_N).$$

That is, maximize a linear combination of the values of the state variables at the fixed time t_N, subject to the system dynamics

$$\frac{dx_i}{dt} = g_i(\bar{x}, V); \quad x_i(t_0) = x_{i0} \quad i = 1, 2, \ldots, n.$$

The maximum principle in this case states that the optimal $V(t)$ must maximize for all t, $t_0 \leqslant t \leqslant t_N$ the following:

$$H = \sum_{j=1}^{n} \lambda_j g_j(\bar{x}, V)$$

where

$$\frac{d\lambda_i}{dt} = -\sum_{j=1}^{n} \lambda_j \frac{\partial g_j}{\partial x_i}$$

$$\lambda_i(t_N) = c_i \quad i = 1, 2, \ldots, n.$$

Example 7.10 Develop the maximum principle equations for the tubular reactor problem discussed in example 7.1.
Solution The dynamic equations are

$$\frac{dx_1}{dt} = -k_1(T)x_1 + k_2(T)x_2; \quad x_1(0) = x_{10} \tag{i}$$

$$\frac{dx_2}{dt} = k_1(T)x_1 - [k_2(T) + k_3(T)]x_2; \quad x_2(0) = x_{20}. \tag{ii}$$

The objective function is

$$\text{Max} \quad 0 \; x_1(1) + 1 \; x_2(1).$$

The Hamiltonian is

$$H(x_1, x_2, \lambda_1, \lambda_2, T) = \lambda_1(-k_1(T)x_1 + k_2(T)x_2)$$

$$+ \lambda_2(k_1(T)x_1 - [k_2(T) + k_3(T)]x_2)$$

where

$$\frac{d\lambda_1}{dt} = +k_1\lambda_1 - k_1\lambda_2; \quad \lambda_1(1) = 0 \tag{iii}$$

$$\frac{d\lambda_2}{dt} = -k_2\lambda_1 + (k_2 + k_3)\lambda_2; \quad \lambda_2(1) = 1 \tag{iv}$$

Temperature, T, must be selected to maximize H.

We will return to this example in more detail in the following chapter. For the moment, it should be recognized that the solution to the problem entails the maximization of H together with the solution of equations (i) through (iv).

7.5.2 Equality Constraints on Initial or Terminal State Variables

In the discussion so far, the initial and terminal values of the state variables either have been specified or have not mattered to the optimization problem. In many cases of practical interest, an equality constraint relating the initial or terminal conditions is given. The boundary conditions on the adjoint variables are modified to account for this more general situation.

The problem can be illustrated in terms of two state variables, x_1 and x_2. We seek the maximum of the following integral with t_0 and t_N fixed.

$$z = \int_{t_0}^{t_N} r(x_1, x_2, V)dt$$

subject to

$$\frac{dx_1}{dt} = f_1(x_1, x_2, V)$$

$$\frac{dx_2}{dt} = f_2(x_1, x_2, V)$$

and the conditions

$$g_0[x_1(t_0), x_2(t_0)] = 0$$

$$g_N[x_1(t_N), x_2(t_N)] = 0.$$

The distinction between this problem and one with end-point variables specified is illustrated on Figure 7.4, where the optimal trajectories are plotted on the $x_1 x_2$-plane. The variable t is a parameter along the extremal curves in these cases.

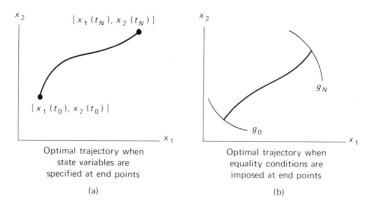

Figure 7.4

Part a of Figure 7.4 shows the case when $[x_1(t_0),\ x_2(t_0)]$ and $[x_1(t_N),\ x_2(t_N)]$ are specified. The optimal trajectory is obtained by solving the Euler-Lagrange equations between the specified boundary points. Part (b) of Figure 7.4 shows the case now under consideration. The optimal trajectory is still obtained by solving the Euler-Lagrange equations. The boundary conditions on these equations now have to do with the way in which the optimal trajectory intersects the constraint curves, g_0 and g_N. The conditions which must be satisfied at these intersections are called *transversality conditions*. They can be obtained by the same principles employed heretofore.

Let $\xi = V - V^*$ represent an arbitrary variation of the decision policy from the optimal. The corresponding variations in x_1 and x_2 are $\epsilon\eta_1 = x_1 - x_1^*$, $\epsilon\eta_2 = x_2 - x_2^*$. The variations are related by the system equations, written to a first-order approximation as

$$\epsilon\dot{\eta}_1 = \left(\frac{\partial f_1}{\partial x_1}\right)_* \epsilon\eta_1 + \left(\frac{\partial f_1}{\partial x_2}\right)_* \epsilon\eta_2 + \left(\frac{\partial f_1}{\partial V}\right)_* \xi$$

$$\epsilon\dot{\eta}_2 = \left(\frac{\partial f_2}{\partial x_1}\right)_* \epsilon\eta_1 + \left(\frac{\partial f_2}{\partial x_2}\right)_* \epsilon\eta_1 + \left(\frac{\partial f_2}{\partial V}\right)_* \xi.$$

Similarly, the performance index variation is

$$
z - z^* = \int_{t_0}^{t_N} \left[\left(\frac{\partial r}{\partial x_1} \right)_* \epsilon \eta_1 + \left(\frac{\partial r}{\partial x_2} \right)_* \epsilon \eta_2 \right.
$$

$$
\left. + \left(\frac{\partial r}{\partial V} \right)_* \xi \right] dt.
$$

In this case we will use the Lagrange multiplier approach to derive the weak form of the maximum principle. A strong form can be derived by the same reasoning as before.

Here, the constraint equations are appended to the integrand of the objective function variation after multiplying each by a Lagrange multiplier. The resultant variation is symbolized by $F - F^*$.

$$
F - F^* = \int_{t_0}^{t_N} \left\{ \left(\frac{\partial r}{\partial x_1} \right)_* \epsilon \eta_1 + \left(\frac{\partial r}{\partial x_2} \right)_* \epsilon \eta_2 \right.
$$

$$
+ \left(\frac{\partial r}{\partial V} \right)_* \xi + \lambda_1 \left[\left(\frac{\partial f_1}{\partial x_1} \right)_* \epsilon \eta_1 + \left(\frac{\partial f_1}{\partial x_2} \right)_* \epsilon \eta_2 \right.
$$

$$
+ \left(\frac{\partial f_1}{\partial V} \right)_* \xi - \epsilon \dot{\eta}_1 \left. \right] + \lambda_2 \left[\left(\frac{\partial f_2}{\partial x_1} \right)_* \epsilon \eta_1 \right.
$$

$$
+ \left(\frac{\partial f_2}{\partial x_2} \right)_* \epsilon \eta_2 + \left(\frac{\partial f_2}{\partial V} \right)_* \xi - \epsilon \dot{\eta}_2 \left. \right] \right\} dt.
$$

By collecting terms, we obtain

$$
F - F^* = \epsilon \int_{t_0}^{t_N} \left[\left(\frac{\partial r}{\partial x_1} + \lambda_1 \frac{\partial f_1}{\partial x_1} + \lambda_2 \frac{\partial f_2}{\partial x_1} \right)_* \eta_1 \right.
$$

$$
\left. - \lambda_1 \dot{\eta}_1 \right] dt + \epsilon \int_{t_0}^{t_N} \left[\left(\frac{\partial r}{\partial x_2} + \lambda_1 \frac{\partial f_1}{\partial x_2} \right. \right.
$$

$$
\left. \left. + \lambda_2 \frac{\partial f_2}{\partial x_2} \right)_* \eta_2 - \lambda_2 \dot{\eta}_2 \right] dt + \int_{t_0}^{t_N} \left[\left(\frac{\partial r}{\partial V} \right. \right.
$$

$$
\left. \left. + \lambda_1 \frac{\partial f_1}{\partial V} + \lambda_2 \frac{\partial f_2}{\partial V} \right)_* \right] \xi \ dt.
$$

Since η_1, η_2 and ξ are arbitrary, the square bracketed terms in each integral must vanish. If we define a Hamilton, H,

$$
H = r + \lambda_1 f_1 + \lambda_2 f_2;
$$

the vanishing of the bracketed term in the last integral means

$$
\left(\frac{\partial H}{\partial V} \right)_* = 0
$$

which is a necessary condition as before.

Integration by parts of the first two integrals yields the adjoint equations, as expected:

$$
\frac{d\lambda_1}{dt} = - \frac{\partial H}{\partial x_1} = - \frac{\partial r}{\partial x_1} - \lambda_1 \frac{\partial f_1}{\partial x_1} - \lambda_2 \frac{\partial f_2}{\partial x_1}
$$

$$\frac{d\lambda_2}{dt} = -\frac{\partial H}{\partial x_2} = -\frac{\partial r}{\partial x_2} - \lambda_1 \frac{\partial f_1}{\partial x_2} - \lambda_2 \frac{\partial f_2}{\partial x_2}.$$

In addition, the integrated portions of the two integrals must vanish

$$\lambda_1(t_0)\, \eta_1(t_0) - \lambda_1(t_N)\, \eta_1(t_N) + \lambda_2(t_0)\, \eta_2(t_0)$$

$$- \lambda_2(t_N)\, \eta_2(t_N) = 0. \tag{7.69}$$

The initial and terminal variations must satisfy the specified equalities. Expansion of these about the optimal values gives

$$g_0\,[x_1(t_0),\ x_2(t_0)] - g_0\,[x_1^*(t_0),\ x_2^*(t_0)]$$

$$= \epsilon\left[\frac{\partial g_0}{\partial x_1(t_0)}\ \eta_1(t_0) + \frac{\partial g_0}{\partial x_2(t_0)}\ \eta_2(t_0)\right].$$

and

$$g_N\,[x_1(t_N),\ x_2(t_N)] - g_N\,[x_1^*(t_N),\ x_2^*(t_N)]$$

$$= \epsilon\left[\frac{\partial g_N}{\partial x_1(t_N)}\ \eta_1(t_N) + \frac{\partial g_N}{\partial x_2(t_N)}\ \eta_2(t_N)\right].$$

Since g_0 and g_N must equal zero for both the optimal and competing values of the state variable, these two equations also equal zero, so that

$$\frac{\partial g_0}{\partial x_1(t_0)}\ \eta_1(t_0) + \frac{\partial g_0}{\partial x_2(t_0)}\ \eta_2(t_0) = 0$$

$$\frac{\partial g_N}{\partial x_1(t_N)}\ \eta_1(t_N) + \frac{\partial g_N}{\partial x_2(t_N)}\ \eta_2(t_N) = 0.$$

Combination of these two equations with (7.69) yields the transversality conditions

$$\lambda_1(t_0) \frac{\partial g_0}{\partial x_2(t_0)} - \lambda_2(t_0) \frac{\partial g_0}{\partial x_1(t_0)} = 0$$

$$\lambda_1(t_N) \frac{\partial g_N}{\partial x_2(t_N)} - \lambda_2(t_N) \frac{\partial g_N}{\partial x_1(t_N)} = 0. \quad (7.70)$$

Equations (7.70) are specific statements of a more general transversality condition. If the problem is stated in terms of n independent variables, x_1, x_2, . . . , x_n subject to m terminal (or initial) constraints

$$g_i[x_1(t_N), x_2(t_N), \ldots, x_n(t_N)] = 0$$

$$i = 1, 2, \ldots, m.$$

The transversality conditions can be obtained by appending these equations to the objective function, premultiplied by suitable Lagrange multipliers. The terminal conditions on the n adjoint variables are then

$$\lambda_i(t_N) = \sum_{k=1}^{m} \mu_k \frac{\partial g_k}{\partial x_i(t_N)} \quad i = 1, 2, \ldots, n \quad (7.71)$$

where the $\{\mu_k\}$ are the (constant) Lagrange multipliers which must be selected in the solution of the problem so that the terminal constraints are satisfied.

Example 7.11 Consider the following dynamic system (the double integrator plant)

$$\frac{dx_1}{dt} = x_2 \quad (i)$$

$$\frac{dx_2}{dt} = V. \tag{ii}$$

If the initial condition $t_0 = 0$ on x_1 and x_2 is specified

$$x_1(0) = 0; \quad x_2(0) = 0. \tag{iii}$$

Find $V(t)$, $0 \leqslant t \leqslant 1$ which minimizes the integral

$$\text{Min } z = \frac{1}{2} \int_0^1 \left\{ [x_1(t)]^2 + [x_2(t)]^2 + V^2 \right\} dt$$

while satisfying the terminal constraint

$$[x_1(1)]^2 + [x_2(1)]^2 = 1. \tag{iv}$$

Solution The problem situation is depicted in the phase plane plot on Figure 7.5. The desired trajectory connects the origin with the unit circle.

We form the Hamiltonian as follows:

$$H = \frac{1}{2} (x_1^2 + x_2^2 + V^2) + \lambda_1 x_2 + \lambda_2 V.$$

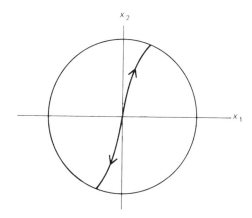

Figure 7.5

The adjoint equations are

$$\frac{d\lambda_1}{dt} = -x_1 \tag{v}$$

$$\frac{d\lambda_2}{dt} = -x_2 - \lambda_1 \tag{vi}$$

and the terminal transversality condition is

$$\lambda_1(1)\ x_2(1) - \lambda_2(1)\ x_1(1) = 0. \tag{vii}$$

The solution to this problem is obtained by solving the differential equations (i), (ii), (v) and (vi) subject to (iii), (iv), and (vii) and the minimizing condition on the Hamiltonian:

$$V = -\lambda_2.$$

Since the differential equations are linear, it is possible to express the solution which satisfies the specified initial conditions in terms of two parameters, λ_{10} and λ_{20}, which are the initial values of $\lambda_1(t)$ and $\lambda_2(t)$ respectively.

$$x_1(t) = \frac{1}{\tau^2 + 1} \left\{ \sqrt{\tau}\, \lambda_{10} \left[\sinh\ (\sqrt{\tau}\ t) \right.\right.$$

$$\left.\left. - \tau \sin\left(\frac{t}{\sqrt{\tau}}\right) \right] - \tau\, \lambda_{20} \left[\cosh\ (\sqrt{\tau}\ t) \right.\right.$$

$$\left.\left. - \cos\left(\frac{t}{\sqrt{\tau}}\right) \right] \right\}$$

$$x_2(t) = \frac{1}{\tau^2 + 1} \left\{ \tau\, \lambda_{10} \left[\cosh\ (\sqrt{\tau}\ t) - \cos\left(\frac{t}{\sqrt{\tau}}\right) \right] \right.$$

$$- \sqrt{\tau} \, \lambda_{20} \left[\tau \, \sinh \, (\sqrt{\tau} \, t) + \sin \left(\frac{t}{\sqrt{\tau}} \right) \right] \Bigg\}$$

$$\lambda_1(t) = 2\lambda_{10} + \frac{1}{\tau^2 + 1} \Bigg\{ - \lambda_{10} \left[\cosh \, (\sqrt{\tau} \, t) \right.$$

$$+ \tau^2 \, \cos \left(\frac{t}{\sqrt{\tau}} \right) \Bigg] + \sqrt{\tau} \, \lambda_{20} \left[\sinh \, (\sqrt{\tau} \, t) \right.$$

$$- \tau \, \sin \left(\frac{t}{\sqrt{\tau}} \right) \Bigg] \Bigg\}$$

$$\lambda_2(t) = \frac{1}{\tau^2 + 1} \Bigg\{ \lambda_{20} \left[\tau^2 \, \cosh \, (\sqrt{\tau} \, t) + \cos \left(\frac{t}{\sqrt{t}} \right) \right]$$

$$- \sqrt{\tau} \, \lambda_{10} \left[\tau \, \sinh \, (\sqrt{\tau} \, t) + \sin \left(\frac{t}{\sqrt{\tau}} \right) \right] \Bigg\}$$

where τ is the positive root satisfying

$$\tau^2 - \tau - 1 = 0.$$

This is the equation from the golden mean search routine; $\tau = 1.618$.

With explicit formulae for x_1, x_2, λ_1 and λ_2 available, the integration constants λ_{10} and λ_{20} can be selected to satisfy conditions (iv) and (vii). Although this chore is cumbersome, it is not too difficult. The values of λ_{10} and λ_{20} which satisfy the terminal constraints are

$$\lambda_{10} = \pm \, 0.61 \quad \lambda_{20} = \pm \, 1.06.$$

The plus and minus signs indicate that the solution is not unique and the decision policy $V*$ can have either positive or negative values. Insertion of these values into the equations above yields both the optimal policy and $x_1 x_2$ trajectory. This trajectory is plotted in the positive quadrant below. There is a similar optimal trajectory in the third quadrant.

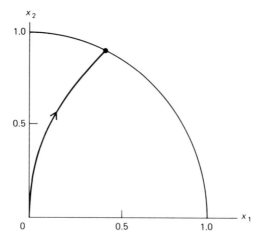

Figure 7.6

7.5.3 Recycle Processes The simplest recycle process is shown in Figure 7.7. Let the dynamics of the process be given by

$$\frac{dx}{dt} = f(x, V) \qquad (7.72)$$

and the equation describing the combining point be

$$x(t_0) = g[x(t_N), x_f] \qquad (7.73)$$

where x_f is a given condition.

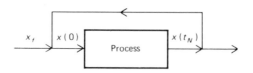

Figure 7.7

The trajectory optimization problem is to select $V(t)$, $t_0 \leqslant t \leqslant t_N$ to maximize z

$$z = \int_{t_0}^{t_n} r(x, V) dt$$

while satisfying the conditions (7.72) and (7.73).

The problem posed here resembles the one previously considered where initial and/or terminal equality conditions were imposed on the state variables. In this case, the recycle stream provides an equality constraint between the initial and terminal values of the state variable. The necessary conditions for this problem can be derived in their weak form by a simple extension of the ideas we have been using. Thus the difference between an arbitrary policy V and the optimal one, V^*, we call ξ; $\epsilon \eta$ is the accompanying difference in state trajectories. Hence, to a first-order approximation

$$z^* - z = \int_{t_0}^{t_N} [r(x^*, V^*) - r(x, V)] \, dt$$

$$z^* - z = \int_{t_0}^{t_N} \left[\left(\frac{\partial R}{\partial x} \right)_* \epsilon \eta + \left(\frac{\partial r}{\partial V} \right)_* \xi \right] dt. \quad (7.74)$$

The variations $\epsilon \eta$ and ξ are related by (7.72)

$$\epsilon \frac{d\eta}{dt} = \left(\frac{\partial f}{\partial x} \right)_* \epsilon \eta + \left(\frac{\partial f}{\partial V} \right)_* \xi. \quad (7.75)$$

Once again requiring that $(\partial f / \partial V)_* \neq 0$, we solve (7.75) for ξ:

$$\xi = \frac{\epsilon \dfrac{d\eta}{dt} - \left(\dfrac{\partial f}{\partial x}\right)_* \epsilon\eta}{\left(\dfrac{\partial f}{\partial V}\right)_*}. \tag{7.76}$$

Substitution of (7.76) into (7.74) gives

$$z^* - z = \epsilon \int_{t_0}^{t_N} \left\{ \left[\left(\frac{\partial r}{\partial x}\right)_* + \lambda \left(\frac{\partial f}{\partial x}\right)_* \right] \eta - \epsilon\lambda \, \dot{\eta} \right\} \, dt$$

where again we use

$$\lambda = - \frac{\left(\dfrac{\partial r}{\partial V}\right)_*}{\left(\dfrac{\partial f}{\partial V}\right)_*}.$$

Integration by parts gives

$$z^* - z = \epsilon \int_{t_0}^{t} \left\{ \left(\frac{\partial r}{\partial x} + \lambda \frac{\partial f}{\partial x} \right) + \frac{d\lambda}{dt} \right\} \eta \, dt$$

$$- \epsilon\lambda\eta \left. \right|_{t_0}^{t_N} \tag{7.77}$$

Since ϵ is arbitrary, the term in curved brackets must vanish, yielding the adjoint equation

$$\frac{d\lambda}{dt} = - \frac{\partial r}{\partial x} - \lambda \frac{\partial f}{\partial x}.$$

The boundary conditions on λ are obtained by requiring that the integrated portion of (7.77) vanish. In this case, however, the variations at $t = t_0$ and $t = t_N$ are related by (7.73), which to a first-order approximation is

$$\eta(t_0) = \frac{\partial g}{\partial x(t_N)} \eta(t_N). \qquad (7.78)$$

Substitution of (7.78) into the integrated portion of (7.77) gives

$$\lambda(t_0) \ \eta(t_0) \ - \ \lambda(t_N) \ \eta(t_N)$$

$$= \ - \ \eta(t_N) \ \left[\lambda(t_N) \ - \ \lambda(t_0) \ \frac{\partial g}{\partial x(t_N)} \right].$$

The requirement that this expression vanish provides the boundary condition on the adjoint variable

$$\lambda(t_N) \ - \ \lambda(t_0) \ \frac{\partial g}{\partial x(t_N)} = 0. \qquad (7.79)$$

This procedure extends to problems involving several independent variables and/or more than one recycle stream or combining point equation.

Hence, the necessary conditions for optimal solutions to problems with recycle (and with other types of non-serial structure) are given by the standard maximum principle conditions, except for modifications which must be made to the boundary conditions on the adjoint variables. Fan has made a comprehensive study of the optimization of complex systems by the maximum principle, and the reader is referred to his work for further generalizations and examples (17).

It should be noted that boundary conditions relate the values of λ at two ends of the interval. This fact complicates the procedures used to solve trajectory optimization problems involving recycle streams.

7.6 *APPENDIX:*
MATHEMATICAL THEOREMS FOR
CONTINUOUS FUNCTIONS

In the body of this chapter, we had occasion to employ several manipulations on continuous functions. A brief synopsis of their mathematic basis is provided in this appendix. Proofs will not be supplied, since they can be found in most texts on advanced calculus or mathematics for engineers and scientists.

7.6.1 Mean Value Theorem Let $f(x)$ be a function of x which is continuous on the interval $a \leqslant x \leqslant b$. Then there is some number x^1, $a \leqslant x^1 \leqslant b$ such that

$$\int_a^b f(x)dx = (b - a)\, \bar{f} \qquad (7.80)$$

where $\bar{f} = f(x^1)$.

7.6.2 Implicit Function Theorem Consider a function of three variables

$$g(x, y, z) = 0.$$

Under certain conditions, an equation of this form has the solution

$$z = G(x, y).$$

These conditions are specified by the following theorem: Let $g(x, y, z)$ be a function defined in an open region about the point (x^0, y^0, z^0). Assume that

$$\left(\frac{\partial g}{\partial z}\right)_{(x^0,\, y^0,\, z^0)} \neq 0.$$

Under these conditions, there is another region, R, about (x^0, y^0, z^0) in which we can write

$$z = G(x, y)$$

and G is continuous in R. Moreover, G has continuous first partial derivatives given by

$$\frac{\partial G}{\partial x} = -\frac{\left(\dfrac{\partial g}{\partial x}\right)}{\left(\dfrac{\partial g}{\partial z}\right)} \text{ and } \frac{\partial G}{\partial y} = -\frac{\left(\dfrac{\partial g}{\partial y}\right)}{\left(\dfrac{\partial g}{\partial z}\right)} . \qquad (7.81)$$

7.6.3 Chain Rule Differentiation

Consider a continuous function of three variables, $\omega = f(x, y, z)$. Suppose in turn the variables (x, y, z) depend on another variable t, viz.

$$x = X(t); \quad y = Y(t); \quad z = Z(t).$$

Substitution of these functions into f yields

$$\omega = f[X(t), Y(t), Z(t)].$$

The derivative of ω with respect to t is then given by the formula

$$\frac{d\omega}{dt} = \left(\frac{\partial f}{\partial x}\right)\frac{dX}{dt} + \left(\frac{\partial f}{\partial y}\right)\frac{dY}{dt} + \left(\frac{\partial f}{\partial z}\right)\frac{dZ}{dt} \qquad (7.82)$$

This expression is a specific form of *chain-rule differentiation*. Many generalizations are possible, but the basic idea is expressed by formula (7.82).

Bibliography

7.3 CALCULUS OF VARIATIONS

Many fine texts have been written on variational calculus. Some of these are

1. Bliss, G.A. *Lectures on the Calculus of Variations*, University of Chicago Press, Chicago, 1946.
2. Bolza, Oskar, *Calculus of Variations*, Chelsea Publishing Co., New York.
3. Courant, R. and D. Hilbert, *Methods of Mathematical Physics*, vol. 1, John Wiley & Sons, Inc., New York, 1953.
4. Forsyth, A.R. *Calculus of Variations*, Dover Publications, New York, 1960.
5. Fox, C. *An Introduction to the Calculus of Variations*, Oxford University Press, New York, 1950.
6. Gelfand, I.M. and S.V. Fomin, *Calculus of Variations* (trans. by R.A. Silverman), Prentice-Hall, Englewood Cliffs, New Jersey, 1963.

7.4 CONTINUOUS DYNAMIC PROGRAMMING

In addition to the texts referred to in the last chapter, the following books provide further examples dealing with the application of variational calculus to engineering problems.

7. Hestenes, Magnus R. *Calculus of Variations and Optimal Control Theory*, John Wiley & Sons, New York, 1966.
8. Weinstock, R. *Calculus of Variations with Applications to Physics and Engineering*, McGraw-Hill Book Co., New York, 1952.

Variational principles are covered in the following:

9. Becker, Martin, *The Principles and Applications of Variational Methods*, Research Monograph No. 27,

Massachusetts Institute of Technology Press, Cambridge, Massachusetts, 1964.

10. Mikhlin, S.G. *Variational Methods in Mathematical Physics*, Pergamon Publishing Co., Elmsford, New York, 1964.

11. Vainberg, M.M. *Variational Methods for the Study of Nonlinear Operators*, Holden-Day, Inc., San Francisco, California, 1964.

and connections with variational calculus:

12. Dreyfus, S.E. *Dynamic Programming and the Calculus of Variations*, Academic Press, Inc., New York, 1965.

13. Jacobson, David H. *Differential Dynamic Programming*, American Elsevier Publishing Co., Inc., New York, 1970.

7.5 THE MAXIMUM PRINCIPLE

The authoritative work is by Pontryagin and his students:

14. Pontryagin, L.S., V.G. Boltanyskii, R.V. Gamkrelidze and E.F. Mischenko, *The Mathematical Theory of Optimal Processes* (translated by K.N. Trirogoff), John Wiley & Sons, Inc., New York, 1962.

A very readable account of the method is by Rozonoer:

15. Rozonoer, L.I. Pontryagin Maximum Principle in the Theory of Optimum Systems. I, *Automation Remote Control* 20, pp. 1288-1302, 1959.

16. Rozonoer, L.I. Pontryagin's Maximum Principle in Optimal System Theory. II, *Automation Remote Control*, 20, pp. 1405-1421, 1959.

17. Fan, L.T. The Continuous Maximum Principle, John Wiley & Sons, New York, 1966.

8
Solution Techniques For Trajectory Optimization Problems

8.1 INTRODUCTION

The maximum principle, continuous dynamic programming and the calculus of variations provide the necessary conditions for a trajectory optimization problem. As in the case of parameter optimization, many techniques have been devised for locating extremal paths which satisfy these conditions. These techniques vary in complexity and in their ability to solve the problem. By reference to the maximum principle statement of the necessary conditions for a trajectory optimization problem, we can place some perspective on these techniques. We distinguish three parts of the necessary conditions:

(1) the system and adjoint equations;
(2) the boundary conditions on these equations;   (8.1)
(3) the requirement that the Hamiltonian be maximized.

The solution to the trajectory optimization problem must satisfy all three of these conditions. In general, if the system equations are nonlinear, conditions (8.1) provide a formidable challenge, and solutions must be obtained with recourse to computerized numerical techniques.

Such techniques involve iterative procedures for solving the problem. Generally, the iteration is on one or other of the conditions in (8.1). That is, a guess is made as to the trajectories which satisfy either condition (1), (2) or (3) in (8.1). On the basis of this guess, the equations are solved to satisfy exactly the remaining two conditions. As a result of this solution, an improved estimate for the unsatisfied condition is made and the process repeated. When the condition iterated upon is satisfied to a preassigned level of accuracy, the problem is considered solved.

When the boundary conditions are selected for iterative solution, the technique is referred to as *approximation to the problem*. If the requirement that the Hamiltonian be maximized is iterated on, the technique is known as *approximation to the solution.* If the system and adjoint equations are iterated on, the technique is known as *approximation to the model* or *trajectory approximation.*

Trajectory approximation belongs to a class of solution techniques for variational problems which are known as *direct methods*. Here a suitable approximation to the trajectory optimization problem is formulated in terms of a finite number of parameters, thereby reducing the problem to a parameter optimization problem.

Many solution techniques for trajectory optimization problems have been proposed based on these ideas. In the remaining pages of this book, we will cover several of the more important techniques and present a number of examples which illustrate the advantages and disadvantages of the various approaches.

8.2 INDIRECT METHODS

The basic difficulty associated with variational problems is the solution of the two-point boundary value problem posed by the Euler-Lagrange equations. The indirect methods we will discuss are, in effect, procedures for overcoming this difficulty. Considerable research has gone on in this area in the last decade, and a voluminous literature has grown up on the subject. The following presentation is based on a small sample of this work, and the interested reader is encouraged to peruse some of the other accounts in the references provided at the end of this chapter.

8.2.1 Approximation to the Problem In order to illustrate techniques based on an approximation to the problem, consider specifically the problem involving a single state and decision variable

$$\frac{dx}{dt} = f(x, \ V) \tag{8.2}$$

$$x(t_0) = x_0 \tag{8.3}$$

$$\text{Max} \int_{t_0}^{t_N} r(x, \ V)dt. \tag{8.4}$$

The Hamiltonian is

$$H = r(x, \ V) + \lambda \ f(x, \ V) \tag{8.5}$$

and the adjoint equation is

$$\frac{d\lambda}{dt} = -\frac{\partial r}{\partial x} - \lambda \frac{\partial f}{\partial x} \tag{8.6}$$

$$\lambda(t_N) = 0. \tag{8.7}$$

Because the system and adjoint equations are nonlinear, a numerical technique in general is required for their solution. Numerical techniques for the solution of differential equations involve replacing the differential equations by finite difference approximations and solving these equations successively in a single direction. (A complete discussion of numerical techniques for the solution of differential equations is beyond the scope of this text. However, a brief description of one of the most commonly used procedures, the Runge-Kutta routine, is described in an appendix to this chapter.) Because part of the boundary conditions are specified at each end of the interval, however, it is impossible to initiate a numerical procedure exactly. Hence the following algorithm is proposed.

Step 1. Guess a value for the initial condition of the adjoint variable. Call it λ_0^1.

Step 2. Solve the system equations (8.2) and (8.6) from $t = t_0$ to $t = t_N$ with condition (8.3) and that from Step 1 as boundary conditions. $V(t)$ must be determined by maximizing H as the solution progresses.

Step 3. Terminate the solution at $t = t_N$. Call the value of the adjoint variable so obtained $\lambda^1(t_N)$.

Step 4. Compare $\lambda^1(t_N)$ to zero. If agreement is as close as desired, the problem is solved. If the difference between the computed and desired values of the adjoint variable exceeds a specified tolerance, the initial guesses for λ must be changed and the problem resolved.

The first passage through the algorithm furnishes a solution to the following problem:

$$\text{Max} \quad \int_{t_0}^{t_N} [r(x, \ V) + \lambda^1(t_N) \ f(x, \ V)] \, dt.$$

Each succeeding passage through the algorithm solves another trajectory optimization problem. Unhappily, these are not necessarily the one posed. Hence the descriptive designation for the technique — approximation to the problem.

The tricky part of the algorithm lies in the selection of a new initial guess for the adjoint variable which is more likely to yield the desired terminal value than the old one. Intuitively, one feels that it should be possible to use the difference between the computed terminal value of the adjoint variable and its desired value to update the initial guess. In order to do this, however, it is necessary to know how a change in an initial guess effects the terminal condition. This change in terminal λ with respect to a change in initial λ can be written as a partial derivative

$$\frac{\partial \lambda(t_N)}{\partial \lambda(t_0)} \ .$$

Hence if $\Delta\lambda(t_N)$ represents a finite change in the terminal λ, produced by a finite change in initial λ, $\Delta\lambda(t_0)$, the relationship between them to a first approximation is

$$\Delta\lambda(t_N) = \frac{\partial \lambda(t_N)}{\partial \lambda(t_0)} \ \Delta\lambda(t_0). \tag{8.8}$$

If we treat $\Delta\lambda(t_N)$ as the difference between desired and computed terminal values for the adjoint variable and for the moment consider the partial derivative known, (8.8) yields $\Delta\lambda(t_0)$ directly. The new guess for λ at $t = t_0$ is then λ_0^2:

$$\lambda_0^2 = \lambda_0^1 + \frac{\Delta\lambda(t_N)}{\left(\dfrac{\partial \lambda(t_N)}{\partial \lambda(t_0)}\right)} \ . \tag{8.9}$$

This value is inserted into Step 1 of the algorithm, and the computation process is repeated. The success of this procedure hinges on the ability to assign a value to the partial derivative which appears in (8.8) and (8.9). Basically, this assignment comes as a result of re-solving the system equations several times. Two ways are proposed.

A. Direct Evaluation from System Equations This procedure is equivalent to numerical differentiation of a function. The derivative of a function $f(x)$ may be numerically determined at a point x^1 by evaluating f at x^1 and at $x^1 + dx$ and by using the formula

$$\left(\frac{df}{dx}\right)_{x^1} \approx \frac{f(x^1 + dx) - f(x^1)}{dx}.$$

In our case, since we have already evaluated $\lambda(t_N)$ corresponding to the initial guess $\lambda^1(t_0)$, we can perturb

$\lambda^1(t_0)$ by a small amount, $d\lambda$, and re-solve the system equations. The terminal value of λ in this case may be identified as

$$\lambda(t_N)_{\lambda^1 + d\lambda}$$

The derivative needed in (8.9) is then approximated as

$$\frac{\partial\lambda(t_N)}{\partial\lambda(t_0)} = \frac{\lambda(t_N)_{\lambda^1 + d\lambda} - \lambda^1(t_N)}{d\lambda}.$$

Although simple to apply, this method of obtaining the partial derivative suffers from two drawbacks. First of all, it requires that the complete optimization problem be solved twice before a change in initial conditions is made. Secondly, the obtaining of partial derivatives in this way is subject to considerable numerical error. If the perturbation $d\lambda$ is too small, the numerical derivative may reflect only roundoff error; if the perturbation is too large, the numerical derivative may be a poor approximation of the actual derivative.

B. Linear Approximation An alternative approach is to use a linearized version of the system equations to obtain the required partial derivatives. This linearization includes both the state and adjoint variables and the decision variables. In this way, the decision policy used in the first passage through the algorithm serves as a basis for estimating policies corresponding to perturbations in initial conditions. This scheme circumvents solving the total problem to obtain the partial derivatives.

Linearization of the state equation produces the following, in terms of the perturbations Δx and ΔV:

$$\Delta\dot{x} = \left(\frac{\partial f}{\partial x}\right)\Delta x + \left(\frac{\partial f}{\partial V}\right)\Delta V. \qquad (8.10)$$

Similarly, for the adjoint equation

$$\Delta\dot{\lambda} = -\left(\frac{\partial^2 r}{\partial x^2} + \lambda\,\frac{\partial^2 f}{\partial x^2}\right)\Delta x - \left(\frac{\partial f}{\partial x}\right)\Delta\lambda$$

$$-\left(\frac{\partial^2 r}{\partial V \partial x} + \lambda\,\frac{\partial^2 f}{\partial V \partial x}\right)\Delta V. \tag{8.11}$$

Although formidable in appearance, the coefficients of each of the perturbation terms are evaluated in the first passage through the algorithm, and they need not be recomputed in the solution to evaluate the desired partial derivative.

The perburbation in the decision variable may be eliminated by observing that

$$\frac{\partial H}{\partial V} = \frac{\partial r}{\partial V} + \lambda\,\frac{\partial f}{\partial V} = 0.$$

Hence perturbations in this equation must sum to zero

$$\left(\frac{\partial^2 r}{\partial x \partial V} + \lambda\,\frac{\partial^2 f}{\partial x \partial V}\right)\Delta x + \left(\frac{\partial f}{\partial V}\right)\Delta\lambda$$

$$+\left[\frac{\partial^2 r}{\partial V^2} + \lambda\,\frac{\partial^2 f}{\partial V^2}\right]\Delta V = 0. \tag{8.12}$$

The term multiplying ΔV is $\partial^2 H/\partial V^2$, which must be nonzero; hence ΔV may be related to the other perturbations by means of this equation.

This allows the perturbation equations to be written as

$$\Delta\dot{x} = a_{11}(t)\,\Delta x + a_{12}(t)\,\Delta\lambda$$

$$\Delta\dot{\lambda} = a_{21}(t)\,\Delta x - a_{11}(t)\,\Delta\lambda \tag{8.13}$$

where

$$a_{11}(t) = \frac{\partial f}{\partial x} - \frac{\dfrac{\partial f}{\partial V}}{\dfrac{\partial^2 H}{\partial V^2}} \left[\frac{\partial^2 r}{\partial x \partial V} + \lambda \frac{\partial^2 f}{\partial x \partial V} \right]$$

$$a_{12}(t) = - \frac{\left(\dfrac{\partial f}{\partial V}\right)^2}{\dfrac{\partial^2 H}{\partial V^2}}$$

$$a_{21}(t) = - \left(\frac{\partial^2 r}{\partial x^2} + \lambda \frac{\partial^2 f}{\partial x^2} \right) + \frac{\left[\dfrac{\partial^2 r}{\partial x \partial V} + \lambda \dfrac{\partial^2 f}{\partial x \partial V} \right]^2}{\dfrac{\partial^2 H}{\partial V^2}} .$$

The $\{a_{ij}\}$ are composed of various partial derivatives that have values obtained in the preceding passage through the algorithm. Consequently, the $\{a_{ij}\}$ will in general be functions of time. Their values will have to be stored in computer memory.

The equation set (8.13) can now be solved with $\Delta x(t_0) = 0$. $\Delta \lambda(t_0)$ is conveniently set equal to one. The resultant value of the terminal adjoint perturbation is the desired partial derivative

$$\frac{\partial \lambda \ (t_N)}{\partial \lambda \ (t_0)} = \frac{\Delta \lambda \ (t_N)}{\Delta \lambda \ (t_0)} = \Delta \lambda \ (t_N).$$

We illustrate the algorithm with the following example.

Example 8.1 Obtain the solution to the optimal control problem stated in example 7.4.

Solution The system equation is

$$\frac{dx}{dt} = - ax + bV \quad x(t_0) = x_0. \tag{i}$$

The performance index is

$$\text{Min} \ \frac{1}{2} \ \int_{t_0}^{t_N} (x^2 + V^2)dt.$$

The Hamiltonian is

$$H = \frac{1}{2} (x^2 + V^2) + \lambda (- ax + bV).$$

The adjoint equation is

$$\frac{d\lambda}{dt} = a\lambda - x; \ \lambda(t_N) = 0. \qquad \text{(ii)}$$

In order for the Hamiltonian to be minimized, the following must hold:

$$V^* = - b\lambda. \qquad \text{(iii)}$$

Hence, in this case, it is important to be able to obtain $\lambda(t)$.

Since equations (i) and (ii) are already linear, it is not necessary to utilize a linearized perturbation equation. Substitution of (iii) into (i) gives

$$\frac{dx}{dt} = - ax - b^2\lambda; \qquad x(t_0) = x_0. \qquad \text{(iv)}$$

Equations (ii) and (iv) were, of course, derived in example 7.4 and their solution obtained by standard means. Here we wish to emphasize the technique just discussed. Hence, let λ_0^1 represent the initial guess for $\lambda(t_0)$. This condition, together with $x(t_0) = x_0$ permits solution of equations (ii) and (iv). Again $m^2 = a^2 + b^2$ and $t_0 = 0$.

$$x(t) = x_0 \left[\cosh \ mt - \frac{a}{m} \ \sinh \ mt \right]$$

$$- \frac{b^2 \lambda_0^1}{m} \ \sinh \ mt \qquad\qquad (v)$$

$$\lambda(t) = - \frac{x_0}{m} \ \sinh \ mt + \lambda_0^1 \left[\cosh \ mt \right.$$

$$\left. + \frac{a}{m} \ \sinh \ mt \right]. \qquad\qquad (vi)$$

Since we require $\lambda(t_N) = 0$, equation (vi) is solved to provide an improved initial guess.

$$\lambda_0^2 = \frac{\dfrac{x_0}{m} \ \sinh \ mt_N}{\left[\cosh \ mt_N + \dfrac{a}{m} \ \sinh \ mt_N \right]}$$

Because the actual system equations are linear, this value of $\lambda(t_0)$ is the exact initial value. This can be demonstrated by insertion of λ_0^2 into equation (v) for λ_0^1. This gives

$$x(t) = x_0 \left[\frac{a \ \sinh \ m(t_N - t) + m \ \cosh \ m(t_N - t)}{a \ \sinh \ m \ t_N + m \ \cosh \ mt_N} \right]$$

which agrees with the solution obtained in example 7.4.

In this case, we were able to use the system equations directly because they were already linear. An analytical solution was possible because the coefficients in the linear equations were constant. In a general case, this latter condition would not hold, and equations (8.13) would require numerical solution. However, once the solution is obtained, it is used to update the initial guess on the adjoint variable in the manner shown above.

As an alternative to the above procedure, one can take advantage of the fact that equations (ii) and (iv) are linear homogeneous differential equations. Their solutions are related as follows

$$\lambda(t) = k(t) \ x(t) \tag{vii}$$

where $k(t)$ is an unknown function of t. To show that (vii) is a solution to the equations, substitute (vii) into equation (ii), and use equation (iv) to replace dx/dt:

$$\frac{dk}{dt} \ x + k[-ax - b^2 kx] = akx - x$$

which, when rearranged, becomes

$$\left[\frac{dk}{dt} - 2ak - b^2 k^2 + 1 \right] x = 0.$$

Since x is not everywhere zero, the bracketed term must equal zero on the interval $[t_0, \ t_N]$. That is,

$$\frac{dk}{dt} = 2ak + b^2 k^2 - 1. \tag{viii}$$

This is a differential equation describing how k varies with t. The boundary condition on k comes from the fact that equation (vii) must hold for all t on the given interval, including $t = t_N$. Here $\lambda = 0$. Since $x(t_N)$ need not be zero, it must hold that

$$k(t_N) = 0. \tag{ix}$$

Hence, it is possible to solve equations (viii) and (ix) for $k(t)$. At $t = t_0$ we can utilize relationship (vii) to obtain

$$\lambda(t_0) = k(t_0) \ x(t_0).$$

In other words, we have used equations (viii) and (ix) to sweep the boundary condition on λ at $t = t_N$ back to $t =$

t_0. This method of solution generalizes to problems with several variables and presents an alternative way of obtaining improved initial guesses for the adjoint variables by means of the linearized perturbation equations.

This idea has further significance in automatic control. If equation (vii) is substituted into equation (iii) which identifies the optimal control policy, we obtain

$$V^* = - bk(t) \ x(t). \tag{x}$$

In other words, equation (x) represents a *feedback control* policy. The optimal V can be obtained by measuring the state variable $x(t)$ and multiplying it by $- bk(t)$. Since $k(t)$ may be entirely precomputed, the feedback control law is independent of initial conditions.

As t_N becomes large, $k(t)$ approaches a constant. Hence $dk(t)/dt = 0$ and we may obtain k by solution of the algebraic equation.

$$2ak + b^2 k^2 - 1 = 0 \ .$$

Whenever the trajectory optimization problem involves more than one state variable, it is necessary to generalize these ideas. Thus, if there are n state variables, a perturbation in the jth adjoint variable at t_N is related to all n of the initial perturbations:

$$\Delta\lambda_j \ (t_N) = \sum_{i=1}^{n} \cdot \frac{\partial\lambda_j \ (t_N)}{\partial\lambda_i \ (t_0)} \ \Delta\lambda_i \ (t_0) \qquad j = 1, \ 2, \ \ldots \ , \ n.$$

In this case, there are n^2 partial derivatives which must be evaluated. This will require either n additional solutions of the optimization problem in order to evaluate directly the derivatives or n solutions of the linearized system equations. In either case, a formidable task.

Once the partial derivatives have been evaluated, the proper changes of the initial guesses can be determined by solving the linear set of equations utilizing Cramer's Rule.

Example 8.2 Obtain the feedback control policy for the system

$$\frac{d^2z}{dt^2} = V$$

which minimizes

$$\frac{1}{2} \int_0^\infty \left[z^2 + a \left(\frac{dz}{dt} \right)^2 + V^2 \right] dt.$$

Solution In order to put the system equation in the proper format, let

$$x_1 = z$$

$$x_2 = \frac{dz}{dt}.$$

The system equations become

$$\frac{dx_1}{dt} = x_2$$

$$\frac{dx_2}{dt} = V.$$

(This change of variables generalizes to n^{th} order differential equations.)

The performance index becomes

$$\frac{1}{2} \int_0^\infty (x_1^2 + ax_2^2 + V^2)dt.$$

The Hamiltonian is defined to be

$$H = \frac{1}{2} (x_1^2 + ax_2^2 + V^2) + \lambda_1 x_2 + \lambda_2 V$$

where

$$\frac{d\lambda_1}{dt} = - x_1 \quad \lambda_1(\infty) = 0$$

$$\frac{d\lambda_2}{dt} = - \lambda_1 - ax_2 \quad \lambda_2(\infty) = 0$$

and

$$V^* = - \lambda_2.$$

In this case, the x's and λ's are related by the more general form

$$\lambda_1 = k_{11} x_1 + k_{12} x_2$$

$$\lambda_2 = k_{12} x_1 + k_{22} x_2$$

and the feedback control law becomes

$$V^* = - k_{12} x_1 - k_{22} x_2.$$

Since the upper limit on the performance index is infinity, the k_{ij} are constant. Substitution of their x-dependent forms into the equations for the k's gives

$$k_{12}^2 = 1$$

$$- k_{11} + k_{12} k_{22} = 0$$

$$- 2k_{12} + k_{22}^2 - a = 0$$

whose solution is

$$k_{12} = + 1$$

$$k_{22} = \sqrt{a + 2}$$

$$k_{11} = \sqrt{a + 2}$$

and

$$V^* = - (x_1 + \sqrt{a + 2} \; x_2).$$

It is possible to generalize the arguments made in the last two examples. This generalization has been made for a large number of systems and constitutes an important area in modern control theory. Those interested in this subject will find that the references listed at the end of the chapter cover this topic in great detail.

8.2.2 Approximation to Solution The basic philosophy of techniques which approximate the solution is the utilization of iterative algorithms to successively improve the decision policy. In a sense, they work on the same premise as optimum seeking parameter optimization techniques except that they must deal with a continuous variable. This produces some complications in practice which can be overcome by using more sophisticated search routines.

In terms of the variational problem posed in section 8.2.1, the computational algorithm used by approximation-to-the-solution methods is as follows.

Step 1. Guess an initial decision policy. Call this $V^1(t)$. Often $V^1(t)$ is chosen as a constant.

Step 2. The system and adjoint equations are solved with the decision policy given by $V^1(t)$. Generally it is preferable to solve the system equations forward from t_0 to t_N and the adjoint equations backward from t_N to t_0. This avoids numerical stability problems which may occur if the adjoint equations are solved forward.

Step 3. With $x(t), \lambda(t)$ and $V(t)$ known, the derivative of H with respect to V is computed for all t, $t_0 \leq t \leq t_N$: $\partial H(t)/\partial V$.

Step 4. If this derivative vanishes for all t, $t_0 \leq t \leq t_N$, the problem is solved. If it does not, the decision policy is updated and the algorithm returns to step 1.

It is the updating of the decision policy which is the tricky part of this algorithm. Several ways of accomplishing this updating have been suggested. As might be suspected, they utilize the nonvanishing derivative $\partial H/\partial V$ to generate an improved decision policy. By analogy to similar parameter optimization situations, we distinguish between policy-improvement techniques based on gradient methods and on second-order methods.

Gradient Methods Since the derivative of H with respect to V is computed over the interval $[t_0, \ t_N]$, we can use this derivative to determine the direction of maximum increase or decrease of H. The decision policy can then be improved by changing $V^1(t)$ for all t, $t_0 \leq t \leq t_N$, in the direction indicated by $\partial H/\partial V$.

Unlike the parameter optimization case, it is not possible to explore conveniently the behavior of H in the direction $\partial H/\partial V$. Hence a definite step size in this direction must be specified. There is a good deal of flexibility in the specification of this step size and no hard and fast rule can be given. If $\partial^2 H/\partial V^2$ is also computed and is negative for all t, $t_0 \leq t \leq t_N$ as it should be, one suggested procedure for modifying the decision policy is

$$V^2(t) = V^1(t) - \frac{\dfrac{\partial H}{\partial V}}{\epsilon \left(\dfrac{\partial^2 H}{\partial V^2} \right)}$$

where

$$0 < \epsilon \leq 1.$$

The choice of ϵ is a somewhat arbitrary matter. If an initial choice provides a satisfactory agreement, then ϵ may be maintained throughout the search. If the change is too large or small, ϵ should be modified accordingly. It is best to evaluate the objective function directly for each succeeding decision policy in order to determine the magnitude of improvement.

The solution procedure involving the partial derivative of the Hamiltonian extends in a direct manner to situations in which there are more than one decision variable. In this case the gradient direction of the Hamiltonian is evaluated for an assumed policy, and policy improvement are therefore made by adjustments in this direction.

As with gradient methods in parameter optimization, this method is simple to implement. However, it demonstrates slow convergence in the vicinity of the optimum.

To circumvent the poor convergence of gradient methods, an alternative policy improvement is proposed based on second-order considerations.

Second-Order Methods Second-order methods involve elaborate procedures for computing the change in decision policy, ΔV, which will reduce $\partial H / \partial V$ to zero with quadratic convergence. In order to develop the equation for ΔV, we linearize the state and adjoint equations as in section 8.2.1.

$$\Delta \dot{x} = \left(\frac{\partial f}{\partial x} \right)_0 \Delta x + \left(\frac{\partial f}{\partial V} \right)_0 \Delta V \qquad (8.14)$$

$$\Delta \dot{\lambda} = - \left(\frac{\partial^2 H}{\partial x^2} \right)_0 \Delta x - \left(\frac{\partial f}{\partial x} \right)_0 \Delta \lambda - \left(\frac{\partial^2 H}{\partial V \partial x} \right)_0 \Delta V \qquad (8.15)$$

where the subscript o has been added to all the partial derivatives to indicate they are evaluated in the first pass through the algorithm using the original guess for the decision policy. In this case, however, the decision policy does not make $\partial H/\partial V$ zero:

$$\frac{\partial H}{\partial V} = \frac{\partial r}{\partial V} + \lambda \frac{\partial f}{\partial V} \neq 0. \qquad (8.16)$$

Hence, to a first-order approximation:

$$\Delta \left(\frac{\partial H}{\partial V} \right) = \left(\frac{\partial^2 H}{\partial x \partial V} \right)_0 \Delta x + \left(\frac{\partial f}{\partial V} \right)_0 \Delta \lambda + \left(\frac{\partial^2 H}{\partial V^2} \right)_0 \Delta V. \qquad (8.17)$$

If we require that the perturbations represent the conditions resulting from the new policy minus those obtained from the old policy, then

$$\Delta \left(\frac{\partial H}{\partial V} \right) = 0 - \left(\frac{\partial H}{\partial V} \right)_0. \qquad (8.18)$$

Substitution of (8.18) into (8.17) and rearrangement gives for ΔV

$$\Delta V = \frac{-1}{\left(\dfrac{\partial^2 H}{\partial V^2} \right)_0} \left[\left(\frac{\partial H}{\partial V} \right)_0 + \left(\frac{\partial^2 H}{\partial x \partial V} \right)_0 \Delta x + \left(\frac{\partial f}{\partial V} \right)_0 \Delta \lambda \right]. \qquad (8.19)$$

For this solution to be valid, we require $\partial^2 H/\partial V^2 < 0$ for all t, $t_0 \leqslant t \leqslant t_N$. The initial guess for $V(t)$ must provide this condition. If it does not, another guess must be made — perhaps by performing a gradient solution on the problem.

Notice that the first term in (8.19) is identical with that suggested by the gradient approach. However, it includes the effects on x and λ which accompany the change ΔV. These effects can be obtained by inserting (8.19) into (8.14) and (8.15):

$$\Delta \dot{x} = a_{11}(t)\Delta x + a_{12}(t)\Delta \lambda + \omega_1(t) \qquad (8.20)$$

$$\Delta \dot{\lambda} = a_{21}(t)\Delta x - a_{11}(t)\Delta \lambda + \omega_2(t) \qquad (8.21)$$

where $a_{11}(t)$, $a_{12}(t)$ and $a_{21}(t)$ have the same meaning they had in the previous section and

$$\omega_1(t) = -\frac{\left(\dfrac{\partial f}{\partial V}\right)_0 \left(\dfrac{\partial H}{\partial V}\right)_0}{\left(\dfrac{\partial^2 H}{\partial V^2}\right)_0} \qquad \omega_2(t) = \frac{\left(\dfrac{\partial^2 H}{\partial V \partial x}\right)_0 \left(\dfrac{\partial H}{\partial V}\right)_0}{\left(\dfrac{\partial^2 H}{\partial V^2}\right)_0}.$$

Equation (8.20) is solved for $\Delta x(t_0) = 0$ and (8.21) with $\Delta \lambda(t_N) = 0$. Unlike the equations which were solved in the approximation-to-the-problem method, these equations are not homogeneous but have forcing functions in terms of $(\partial H/\partial V)_0$. Moreover, their solutions are required over the entire t-interval as required by (8.19).

As an alternative to solving (8.20) and (8.21), the sweep method used earlier can be applied. Here Δx and $\Delta \lambda$ are related by

$$\Delta \lambda = k \, \Delta x + h \qquad (8.22)$$

where k and h are unknown functions of t. Because $\Delta \lambda(t_N) = 0$, $k(t_N)$ and $h(t_N)$ must equal zero. Substitution of (8.22) into (8.20) and (8.21) shows that k and h must satisfy the following equations:

$$\dot{k} = -2a_{11}k - a_{12}k^2 + a_{21}; \qquad k(t_N) = 0 \qquad (8.23)$$

$$\dot{h} = -a_{11}h - a_{12}kh - k\,\omega_1(t) + \omega_2(t); \qquad h(t_N) = 0.$$

$$(8.24)$$

Equations (8.23) and (8.24) can be solved backwards from t_N with no knowledge of Δx or $\Delta \lambda$. Then (8.20) can be solved forward from t_0 in the modified form:

$$\Delta \dot{x} = a_{11}(t)\Delta x + a_{12}(t)\,[k\Delta x + h] + \omega_1(t).$$

The results are then substituted into (8.19) to improve the decision policy.

The general experience with second-order methods is that the increased computations required over and above first-order methods are more than compensated by the increased convergence of the routine.

The policy-improvement techniques presented can be generalized to problems with many state and decision variables. In addition, they can be applied to problems in which the final time is not specified. Constraints on the decision variables can also be handled. A particularly lucid account of these important general cases is presented in the text by Bryson and Ho (2).

Example 8.3   Solve the following trajectory optimization problem by approximation-to-the-solution techniques.

$$\text{Min } \frac{1}{2} \int_0^1 (x^2 + V^2)dt$$

subject to

$$\frac{dx}{dt} = -Vx \quad x(0) = 1$$

Solution We first solve the problem analytically. Form the Hamiltonian

$$H = \frac{1}{2} (x^2 + V^2) - \lambda(Vx) \qquad \text{(i)}$$

where

$$\frac{d\lambda}{dt} = -x + \lambda V \qquad \text{(ii)}$$

$$\lambda(1) = 0. \qquad \text{(iii)}$$

In order for the Hamiltonian to be minimized, it is necessary that

$$\frac{\partial H}{\partial V} = V - \lambda x = 0.$$

Hence

$$V^* = \lambda x.$$

Thus, an exact solution is obtained by solving

$$\frac{dx}{dt} = -\lambda x^2 \quad x(o) = 1 \qquad \text{(iv)}$$

$$\frac{d\lambda}{dt} = -x(1 - \lambda^2) \quad \lambda(1) = 0. \qquad \text{(v)}$$

The solution is facilitated by dividing equation (v) by (iv)

$$\frac{d\lambda}{dx} = \frac{(1 - \lambda^2)}{\lambda x}$$

or

$$\int \frac{-2\lambda d\lambda}{1 - \lambda^2} = \int \frac{-2dx}{x}.$$

This equation can be integrated directly to give

$$1 - \lambda^2 = \frac{c^2}{x^2}$$

where c is the constant of integration. In order for the boundary condition on λ to be satisfied, c must equal $x(1)$.

By solving for λ, we obtain

$$\lambda x = \sqrt{x^2 - c^2}.$$

Substitution into the state equation gives

$$\frac{dx}{dt} = -x\sqrt{x^2 - c^2}$$

which can be separated as follows

$$\int_1^x \frac{dx}{x\sqrt{x^2 - c^2}} = -\int_0^t dt.$$

After carrying out the integration,

$$\frac{1}{c}\left[\cos^{-1}\left(\frac{c}{x}\right) - \cos^{-1}(c)\right] = -t. \qquad \text{(vi)}$$

At $t = 1$, $c/x = 1$, since $c = x(1)$. Since $\cos^{-1}(1) = 0$, we find

$$\frac{1}{c}\cos^{-1} c = 1$$

or

$$\cos c = c$$

This is satisfied for $c = 0.74$. Hence,

$$V* = \sqrt{x^2 - (0.74)^2} \qquad \text{(vii)}$$

where x is obtained from equation (vi). A more convenient way of expressing (vi) is

$$x = \frac{c}{\cos [c (1 - t)]}.$$

A plot of $V*$ versus t is given in the figure below.

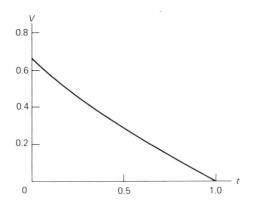

Figure 8.1

We now seek a solution by a gradient method. In order to initiate the algorithm, we assume an arbitrary initial decision policy.

$$V^1(t) = \frac{1}{2} \quad 0 \leqslant t \leqslant 1$$

It is possible to integrate the state equation

$$\frac{dx}{dt} = - \frac{1}{2} x.$$

Hence,

$$x = e^{-\frac{1}{2}t}$$

is the solution which satisfies the boundary condition. The adjoint equation is now

$$\frac{d\lambda}{dt} = -x + \frac{1}{2}\lambda = e^{-\frac{1}{2}t} + \frac{1}{2}\lambda.$$

The solution to this equation, satisfying the adjoint boundary condition, is

$$\lambda = e^{\frac{1}{2}t} \quad (e^{-t} - e^{-1}).$$

The gradient of H is now

$$\frac{\partial H}{\partial V} = V - \lambda x = \frac{1}{2} - (e^{-t} - e^{-1})$$

$$\frac{\partial H}{\partial V} = 0.866 - e^{-t}.$$

Since $\partial^2 H/\partial V^2 = 1$, a suitable normalizing factor is e. Hence,

$$V^2(t) = \frac{1}{2} - \frac{1}{\epsilon}(0.866 - e^{-t})$$

where the minus sign is included since we seek the minimum of the Hamiltonian. For convenience, take $\epsilon = 1$, then

$$V^2(t) = e^{-t} - 0.366.$$

It is still possible to solve the state differential equation analytically with this policy:

$$\frac{dx}{dt} = - x \ (e^{-t} - 0.366).$$

The solution is

$$\ell\eta \ x = (e^{-t} - 1) + 0.366t, \qquad \text{(viii)}$$

but the adjoint equation becomes quite complex:

$$\frac{d\lambda}{dt} = \lambda \ [e^{-t} - 0.366] - \exp \ [0.366t + e^{-t} - 1].$$

In an actual gradient solution, this equation would be integrated numerically to obtain $\lambda(t)$ and the gradient of H evaluated as before. In order to illustrate the results without recourse to a computer, we note that

$$\frac{d(\lambda x)}{dt} = \lambda \ \frac{dx}{dt} + x \ \frac{d\lambda}{dt}.$$

Substitution into this expression from the state and adjoint differential equation gives

$$\frac{d(\lambda x)}{dt} = \lambda \ (- Vx) + x \ (- x + \lambda V) = - x^2.$$

Hence, we can obtain λx by integration

$$\int_0^{\lambda x} d(\lambda x) = - \int_1^t x^2 \ dt. \qquad \text{(ix)}$$

We can obtain x from equation (viii) and integrate the right-hand integral by Simpson's rule. A plot of x versus t is given below. By using values of x from this plot, we obtain the values for λx listed in Table 8.1.

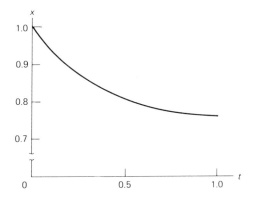

Figure 8.2

Table 8.1. Values of $V^2(t)$ and $\lambda x(t)$

t	$V^2(t)$	λx	$\partial H/\partial V$
0.0	0.63	0.69	−0.06
0.2	0.45	0.52	−0.07
0.4	0.30	0.37	−0.07
0.6	0.18	0.24	−0.06
0.8	0.08	0.12	−0.04
1.0	0.00	0.00	−0.00

The new policy $V^3(t)$ is just λx if we take $\epsilon = 1$ as in the previous formula. The progression of the algorithm is displayed on Figure 8.3 below. Since $V^3(t)$ is almost identical to V^*, we will not carry the gradient technique further.

It is informative to apply the second-order technique to this problem. Formally, the second-order algorithm begins identically to the gradient method; an initial decision policy is selected, λ and x evaluated and $\partial H/\partial V$ obtained.

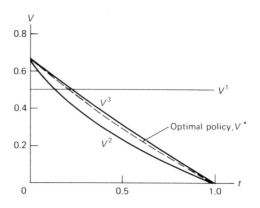

Figure 8.3

We then apply (8.19)

$$\Delta V = - \frac{1}{\left(\dfrac{\partial^2 H}{\partial V^2} \right)} \left[\left(\frac{\partial H}{\partial V} \right)_0 + \left(\frac{\partial^2 H}{\partial x \, \partial V} \right)_0 \Delta x + \left(\frac{\partial f}{\partial V} \right)_0 \Delta \lambda \right].$$

(x)

Here

$$\left(\frac{\partial H}{\partial V} \right)_0 = 0.866 - e^{-t}$$

$$\left(\frac{\partial^2 H}{\partial x \, \partial V} \right)_0 = -\lambda = -e^{-\frac{1}{2}t} + e^{-(1 - \frac{1}{2}t)}$$

$$\left(\frac{\partial f}{\partial V}\right)_0 = - x = - e^{-\frac{1}{2}t}.$$

The perturbations Δx and $\Delta \lambda$ are obtained by solving equations (8.20) and (8.21):

$$\Delta \dot{x} = a_{11}(t)\Delta x + a_{12}(t)\Delta \lambda + \omega_1(t)$$

$$\Delta \dot{\lambda} = a_{21}(t)\Delta x - a_{11}(t)\Delta \lambda + \omega_2(t).$$

Here

$$a_{11}(t) = \frac{\partial f}{\partial x} - \frac{\left(\dfrac{\partial f}{\partial V}\right)}{\dfrac{\partial^2 H}{\partial V^2}}\left[\frac{\partial^2 r}{\partial x \partial V} + \lambda \frac{\partial^2 f}{\partial x \partial V}\right]$$

$$a_{11}(t) = - V - \frac{(-x)}{1}\left[0 + \lambda(-1)\right] = - V - \lambda x$$

$$a_{12}(t) = - \frac{\left(\dfrac{\partial f}{\partial V}\right)^2}{\left(\dfrac{\partial^2 H}{\partial V^2}\right)} = - (-x)^2$$

$$a_{21}(t) = -\left(\frac{\partial^2 r}{\partial x^2} + \lambda \frac{\partial^2 f}{\partial x^2}\right) + \frac{\left(\dfrac{\partial^2 r}{\partial x \partial V} + \lambda \dfrac{\partial^2 f}{\partial x \partial V}\right)^2}{\dfrac{\partial^2 H}{\partial V^2}}$$

$$a_{21}(t) = -\left(1 + \lambda \cdot (0)\right) + \left(0 + \lambda(-1)\right)^2 = -1 + \lambda^2.$$

The forcing functions, ω_1 and ω_2 are

$$\omega_1 = -\frac{\left(\dfrac{\partial f}{\partial V}\right)_0 \left(\dfrac{\partial H}{\partial V}\right)_0}{\left(\dfrac{\partial^2 H}{\partial V^2}\right)_0} = -(-x)(V - \lambda x)$$

$$\omega_2 = \frac{\left(\dfrac{\partial^2 H}{\partial V \partial x}\right)_0 \left(\dfrac{\partial H}{\partial V}\right)_0}{\left(\dfrac{\partial^2 H}{\partial V^2}\right)_0} = (-\lambda)(V - \lambda x).$$

Hence, the equations for Δx and $\Delta\lambda$ become

$$\Delta\dot{x} = -\left(\frac{1}{2} + e^{-t} - e^{-1}\right)\Delta x - e^{-t}\ \Delta\lambda$$

$$+ \ e^{-\frac{1}{2}t}\left(\frac{1}{2} - e^{-t} + e^{-1}\right)$$

$$\Delta\dot{\lambda} = -\left(1 - e^{-t} + 2e^{-1} - e^{-2}\right)\Delta x + \left(\frac{1}{2} + e^{-t} - e^{-1}\right)\Delta\lambda$$

$$- \ e^{\frac{1}{2}t}\left(e^{-t} - e^{-1}\right)\left(\frac{1}{2} - e^{-t} + e^{-1}\right).$$

The complexity of these equations precludes closed form solution here. However, with the aid of a digital computer, they can be solved without much difficulty, although their boundary conditions ($\Delta x(0) = 0$; $\Delta\lambda(1) = 0$) do present a two-point boundary value problem.

These equations were solved numerically on a UNIVAC 1108 by using the Runge-Kutta routine described in the appendix. The results are shown in Figure 8.4 below. Substitution of these values for Δx and $\Delta \lambda$ into equation (x) yields the new policy shown below in Table 8.2.

Table 8.2. New Decision Policy

t	$V^2(t)$
0.0	0.69
0.2	0.52
0.4	0.37
0.6	0.24
0.8	0.12
1.0	0.00

Notice that this policy, obtained from the first pass through the second-order algorithm is virtually identical with that obtained on the second pass through the gradient algorithm. The price for this accelerated convergence is the necessity of using a computer to solve the perturbation equations.

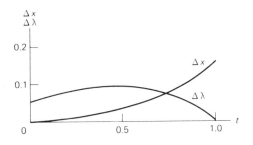

Figure 8.4

8.3 DIRECT METHODS

The solution of trajectory optimization problems by direct methods generally involves an approximation for the solution in terms of a set of unknown coefficients and

specified functions of the independent variable. This reduces the problem to one in terms of the unknown constants, i.e., a parameter optimization problem.

In order to be more specific, consider the problem of
maximizing the following integral

$$\text{Max } z = \int_{t_0}^{t_N} R(x, \dot{x}, t)dt \qquad (8.25)$$

subject to the conditions that

$$x(t_0) = x_0 \qquad (8.26)$$

$$x(t_N) = x_N. \qquad (8.27)$$

Direct methods express the solution to this problem,
$x^*(t)$ as a linear combination of n parameters called mixing
coefficients, $a_1, a_2, \ldots, a_n$ and n specified functions of
t called trial functions, $\phi_1, \phi_2, \ldots, \phi_n$.

$$x^*(t) = \widetilde{x}(t) = \phi_0(t) + \sum_{i=1}^{n} a_i \phi_i(t) \qquad (8.28)$$

where the tilde on $x(t)$ denotes the approximation.

The $\{\phi_i\}$ may be any members of a complete set of
functions on the t interval. For example, they could be
simply powers of t or orthogonal polynomials. The $\{\phi_i\}$
should be selected to give the approximate solution the
general shape of the exact solution, if this is known beforehand. In addition, the function $\phi_0(t)$ should be selected to
satisfy the boundary conditions (8.26) and (8.27), regardless of the values of the mixing coefficients. One example is

$$\widetilde{x}(t) = x_0 \left(\frac{t_N - t}{t_N - t_0} \right) + x_N \left(\frac{t - t_0}{t_N - t_0} \right) + \sum_{i=1}^{n} a_i \phi_i(t)$$

where the remaining $\{\phi_i\}$ must satisfy homogeneous boundary conditions at $t = t_0$ and $t = t_N$.

Substitution of the approximation (8.28) into the integral (8.25) gives the objective function in terms of the mixing coefficients

$$z(a_1, a_2, \ldots, a_n) = \int_{t_0}^{t_N} R(\widetilde{x}(t), \dot{\widetilde{x}}(t), t)dt. \quad (8.29)$$

If the integral can be symbolically integrated, it will yield an expression for z in terms of the $\{a_i\}$. The $\{a_i\}$ are then selected to maximize z.

If the integral cannot be symbolically integrated, we can employ the necessary condition for maximization of z with respect to the $\{a_i\}$, namely that the partial derivatives should vanish.

$$\frac{\partial z}{\partial a_i} = \int_{t_0}^{t_N} \left[\left(\frac{\partial R}{\partial x} \right) \phi_i + \frac{\partial R}{\partial \dot{x}} \dot{\phi}_i \right] dt = 0$$

$$i = 1, 2, \ldots, n. \quad (8.30)$$

This generates n algebraic equations in terms of the n unknown mixing coefficients. The solution of these equations provides the maximizing values for the mixing coefficients, $(a_1^*, a_2^*, \ldots, a_n^*)$ which can be substituted into (8.28) to obtain a near-optimal solution to the problem.

In actual practice, (8.30) is not used to identify the $\{a_i^*\}$, since an easier but more general alternative exists. If the second term in the integral (8.30) is integrated by parts, the following equation results

$$\frac{\partial z}{\partial a_i} = \int_{t_0}^{t_N} \left\{ \left(\frac{\partial R}{\partial x} \right) \phi_i - \frac{d}{dt} \left(\frac{\partial R}{\partial \dot{x}} \right) \phi_i \right\} dt$$

$$+ \left(\frac{\partial R}{\partial \dot{x}} \right) \phi_i \Bigg|_{t_0}^{t_N} = 0.$$

The integrated portion of this equation equals zero since the ϕ_i were chosen to satisfy the homogeneous boundary conditions. Notice that, if the problem specifications on x were made at only one end of the interval, the integrated portion would still vanish, because $(\partial R/\partial \dot{x})$ would vanish at the other end of the interval as a natural boundary condition.

The remaining portion of the integral can be expressed as follows

$$\int_{t_0}^{t_N} \left\{ \frac{\partial R}{\partial x} - \frac{d}{dt} \frac{\partial R}{\partial \dot{x}} \right\} \phi_i(t)dt = 0$$

$$i = 1, 2, \ldots, n. \tag{8.31}$$

The term in curved brackets has the form of the Euler-Lagrange equation. It is not identically zero here, however, because R is evaluated for $\widetilde{x}(t)$ — the approximate solution. It thus represents the *residual*, the failure of the Euler-Lagrange equation to be satisfied by the approximate solution. Equation (8.31) requires that the residual be orthogonal to all of the trial functions $\{\phi_i\}$. If we let ρ represent this residual, we write

$$\int_{t_0}^{t_N} \rho \ \phi_i dt = 0 \quad i = 1, 2, \ldots, n. \tag{8.32}$$

This procedure can be used generally for the approximate solution of differential equations. It is known as *Galerkin's method*. In practice, it is sometimes more efficient to have the residual be orthogonal to a different set of functions than the trial functions. These are called coordinate functions and symbolized by Ψ_i, $i = 1, 2, \ldots, n$. Equation (8.32) is then

$$\int_{t_0}^{t_N} \rho \ \Psi_i \ dt = 0 \quad i = 1, 2, \ldots, n. \tag{8.33}$$

The procedure in this more general sense is referred to as the method of weighted residuals. Two common forms for the coordinate functions or weights are: (1) powers of t in which case the integrals in (8.33) are moments and (2) Dirac delta functions which reduce the integrals to the integrands evaluated at specified points.

Direct methods thus use approximation theory as a substitute for the exact solution of the two-point boundary value problem. In many cases, this is an attractive alternative, especially when the precise shape of the optimal trajectories is not essential. Thus the direct methods of variational calculus are of particular interest to the practicing engineer.

Example 8.4 A Tubular Reactor with Axial Dispersion In order to illustrate how the method of weighted residuals can be used to obtain approximate solutions to differential equations, consider the case of a tubular reactor in which axial dispersion must be taken into account. Consider that a simple first-order reaction is taking place:

$$A \overset{k}{\rightarrow} B.$$

The describing differential equation for this system is

$$\frac{1}{Pe} \frac{d^2x}{dt^2} - \frac{dx}{dt} = k\,x \tag{i}$$

where x is the mole fraction of A, k is the reaction rate constant, Pe is the axial Péclet number and t is the normalized reactor length. If pure A is fed to the reactor, the boundary conditions are

$$x(0) - \frac{1}{Pe} \frac{dx}{dt}\bigg|_{t=0} = 1$$

$$\frac{dx}{dt}\bigg|_{t=1} = 0.$$

An approximating function which satisfies these boundary conditions is

$$\widetilde{x} = 1 + a(t^2 - 2t - 2/Pe)$$

where a is the single mixing coefficient.

In terms of this approximating function, the residual for equation (i) is

$$\rho = \frac{2a}{Pe} - 2a(t - 1) - k \ [1 + a(t^2 - 2t - 2/Pe)].$$

If the trial function is used as the coordinate function, we obtain the following explicit form for a:

$$a = \frac{\displaystyle\int_0^1 k[t^2-2t-2/Pe]\,dt}{\displaystyle\int_0^1 (t^2-2t-2/Pe)\,[2/Pe-2(t-1)-k(t^2-2t-2/Pe)]\,dt}. \text{(ii)}$$

In the case of an isothermal reactor, k and Pe are constants and it is possible to evaluate the integrals in equation (ii) analytically. For the case $Pe = 1$, $k = 2$, we obtain

$$a^* = \frac{160}{667}.$$

It is also possible to solve equation (i) exactly for Pe, k constant. In the case considered, the exact solution is

$$x(t) = \frac{e^{2t} + 2e^{(3-t)}}{4e^3 - 1}.$$

The fit between approximate and exact solution is quite good, and it may be concluded that the trial function selected is a reasonable one. Although this example is a trivial one, it does indicate that naturally occurring two-point boundary value problems can be solved to a good approximation by the method.

A comparison of the exact and approximate solutions is given in Table 8.3

Table 8.3. Comparison of Exact and Approximate Solution

t_0	Exact	Approximate
0.0	0.518	0.518
0.2	0.432	0.430
0.4	0.366	0.363
0.6	0.318	0.315
0.8	0.288	0.286
1.0	0.277	0.277

Example 8.5 Use the direct method to obtain an approximate solution to the classical variational calculus problem of finding the minimal surface of revolution. In this problem, the curve $x(t)$ is rotated about the t-axis. The resulting surface, bounded by the planes $t = t_0$ and $t = t_N$, has area given by

$$\text{Area} = 2\pi \int_{t_0}^{t_N} x \sqrt{1 + (\dot{x})^2} \; dt.$$

To be specific, we will consider $t_0 = 0$ and $t_N = 1$. We also specify that $x(0) = 2$ and $x(1) = 4$. Hence, we seek $x(t)$ for $0 \leqslant t \leqslant 1$ which minimizes the following

$$\text{Min } z = \int_0^1 (x \sqrt{1 + (\dot{x})^2}) \; dt$$

subject to

$$x(0) = 2$$
$$x(1) = 4.$$

Solution We first obtain the exact solution. The integrand of the objective function does not contain t explicitly. In this case, the first integration of the Euler-Lagrange equation can be obtained directly. Note that

$$\frac{d}{dt}\left(\dot{x}\,\frac{\partial R}{\partial \dot{x}} - R\right) = \dot{x}\,\frac{d}{dt}\left(\frac{\partial R}{\partial \dot{x}}\right) + \frac{\partial R}{\partial \dot{x}}\,\ddot{x}$$

$$-\frac{\partial R}{\partial x}\,\dot{x} - \frac{\partial R}{\partial \dot{x}}\,\ddot{x} = \dot{x}\left[\frac{d}{dt}\left(\frac{\partial R}{\partial \dot{x}}\right) - \frac{\partial R}{\partial x}\right] = 0.$$

The last term in square brackets is the Euler-Lagrange equation which must vanish. Since the left-hand side of this equation is a perfect differential, it can be integrated to yield

$$\dot{x}\,\frac{\partial R}{\partial \dot{x}} - R = c,\text{ a constant.}$$

For the example problem, this provides

$$\frac{x}{\sqrt{1 + (\dot{x})^2}} = \frac{1}{c}.$$

The solution to this equation is

$$x = \frac{1}{c}\,\cosh\,(ct + c_1)$$

where c and c_1 are constants which must be selected to fit the boundary conditions. For those stated, we find

$$x = \frac{1}{0.795}\,\cosh\,(0.795t + 1.035).$$

The Euler-Lagrange equation for this problem is

$$x\ddot{x} - [1 + (\dot{x})^2] = 0.$$

As an approximate solution to this equation, we choose the following

$$\widetilde{x}(t) = 2(1 + t) + at(1 - t)$$

where a is the single unknown mixing coefficient. The needed derivatives are

$$\dot{\tilde{x}} = 2 + a(1 - 2t)$$

$$\ddot{\tilde{x}} = - 2a.$$

The residual ρ is

$$\rho = - 2a \ [2 \ (1 + t) + at \ (1 - t)] - 1$$

$$- [2 + a \ (1 - 2t)]^2$$

The Galerkin form of the weighted residual formula is

$$\int_0^1 \rho \ t \ (1 - t) \ dt = 0.$$

Substitution of ρ into the above integral leads to the following equation for a:

$$3a^2 + 30a + 25 = 0.$$

Hence, $a = - 0.9$, and $\tilde{x}(t)$ is

$$\tilde{x}(t) = 2 + 1.1t + 0.9t^2.$$

A comparison of $x^*(t)$ and $\tilde{x}(t)$ is given in Table 8.4 below. The agreement is quite good and again demonstrates that direct techniques can be used to obtain reasonable approximations to variational problems.

Table 8.4. Comparison of Optimal and
Near-optimal Trajectories

t	$x^*(t)$	$\tilde{x}(t)$
0.0	2.00	2.00
0.2	2.26	2.26
0.4	2.58	2.59
0.6	2.99	2.99
0.8	3.46	3.44
1.0	4.00	4.00

Whenever the trajectory optimization problem is phrased explicitly in terms of state and decision variables, direct methods are applied in a different way. Thus, consider the problem posed at the beginning of section (8.2)

$$\text{Max} \int_{t_0}^{t_N} r(x, V)dt \qquad (8.34)$$

subject to

$$\frac{dx}{dt} = f(x, V) \quad x(t_0) = x_0. \qquad (8.35)$$

In this case, there are two possible direct approaches. The first consists of approximating the decision variable, V, in terms of a linear combination of trial functions and mixing coefficients

$$V \approx \widetilde{V} = \sum_{i=1}^{n} a_i \phi_i(t). \qquad (8.36)$$

Substitution of this approximation for V into the differential equation for x – (8.35) should permit x to be obtained as a function of the mixing coefficients. Substitution of the x-value so obtained and V from (8.36) into the objective function reduces it to an expression in terms of the mixing coefficients which can then be selected to achieve the desired maximum. If numerical solution of the differential equation is required, this procedure may involve an iterative search procedure to find the optimizing values of the mixing coefficients.

Some of the difficulties associated with this approach are illustrated in the next example.

Example 8.6   Consider a first-order chemical reaction

$$A \underset{k_2}{\overset{k_1}{\rightleftarrows}} B$$

taking place in a tubular reactor. It is desired to maximize the conversion of A at the reactor outlet. The equation describing the rate of change of the mole fraction of A, x, with normalized reactor length, t is

$$\frac{dx}{dt} = -k_1 x + k_2 (1 - x) \quad 0 \leqslant t \leqslant 1$$

and the inlet composition is

$$x_1(0) = x_0.$$

The analysis is simplified and the essence of the problem is preserved by considering the special case

$$k_2 = k_1^2 = V^2$$

and

$$x_0 = 0.9.$$

Then we find

$$\frac{dx}{dt} = -(V + V^2) x + V^2 \qquad \text{(i)}$$

$$x(0) = 0.9. \qquad \text{(ii)}$$

We now employ a direct method in an effort to solve this problem.

Solution Assume the optimal decision (temperature) policy may be approximated as

$$V = a_0 + a_1 t. \qquad \text{(iii)}$$

With this approximation, it is possible to integrate equation (i) subject to (ii) to obtain

$$x(t) = \exp \left\{ - \left[a_0(1 + a_0)t + \frac{a_1}{2}(1 + 2a_0)t^2 \right. \right.$$

$$\left. \left. + \frac{a_1^2}{3}t^3 \right] \right\} \cdot \left[0.9 + \int_0^t (a_0 + a_1\tau)^2 \exp \right.$$

$$\left. \left\{ a_0(1 + a_0)\tau + \frac{a_1}{2}(1 + 2a_0)\tau^2 + \frac{a_1^2}{3}\tau^3 \right\} d\tau \right].$$

The objective function of maximizing outlet conversion is equivalent to minimizing $x(1)$. Hence the trajectory optimization problem is reduced to one of finding values of a_0 and a_1 which minimize $x(1)$.

It is possible to find a_0^* and a_1^* by one of the unconstrained parameter optimization techniques discussed in chapters four and five. Although the problem is not simple, it is possible to determine the optimizing values as

$$a_0^* = 1.55 \quad a_1^* = -1.125. \tag{iv}$$

The exact solution to this problem can be obtained by application of the maximum principle. The Hamiltonian is

$$H = \lambda \left[V^2 - (V + V^2)x \right]$$

where

$$\frac{d\lambda}{dt} = (V + V^2)\lambda; \quad \lambda(1) = 1$$

and to minimize H

$$V^* = \frac{x}{2(1 - x)}. \tag{v}$$

Since the optimal policy does not depend on λ, we find that

$$\frac{dx}{dt} = \frac{x^2}{4(1-x)^2} - \frac{x^2}{2(1-x)} - \frac{x^3}{4(1-x)^2}$$

$$\frac{dx}{dt} = -\frac{x^2}{4(1-x)}.$$

This equation can be integrated subject to the specified inlet condition to yield

$$\frac{1}{x} + \ell\eta\left(\frac{x}{0.9}\right) = \frac{1}{0.9} + \frac{t}{4}. \qquad \text{(vi)}$$

If the value of x from this expression is substituted into equation (v) for V^*, the optimal decision policy is obtained. Figure 8.5 below plots the optimal and approximate decision policies as a function of t. The approximate policy is obtained by substitution of the optimum mixing coefficient values from equation (iv) into equation (iii).

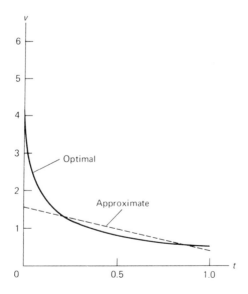

Figure 8.5

Although it is possible to obtain approximate solutions
to trajectory optimization problems in this way, there are
two difficult steps associated with the technique. The first
is the solution of the system differential equations. It may
not always be possible to express the solution symbolically
for a given policy as was done here. In that case, numerical
solution is required, considerably complicating the procedure. Secondly, even if the system differential equations
can be solved, they may involve complicated expressions.
Substitution of these expressions into the objective function may result in a formidable parameter optimization
problem.

An alternative procedure is to approximate the state
variable trajectory in terms of a set of trial functions and
mixing coefficients. By inserting this approximation into
the maximum principle formulation of the problem, it is
possible to utilize the relationships among the state, adjoint
and decision variables to determine values for the mixing
coefficients. This technique has been called trajectory approximation.

8.3.1 Trajectory Approximation The maximum principle formulation of the problem posed by equations
(8.34) and (8.35) consists in forming a Hamiltonian, H

$$H = r(x, V) + \lambda f(x, V) \tag{8.37}$$

where

$$\frac{d\lambda}{dt} = -\frac{\partial r}{\partial x} - \lambda \frac{\partial f}{\partial x} \tag{8.38}$$

$$\lambda(t_N) = 0 \tag{8.39}$$

and H must be maximized by V^*.

Trajectory approximation approximates both the state
and adjoint variables of the problem in terms of a linear
combination of trial functions and mixing coefficients, viz

$$x \approx \tilde{x} = \phi_0(t) + \sum_{i=1}^{n} a_i \phi_i(t) \qquad (8.40)$$

$$\lambda \approx \tilde{\lambda} = \chi_0(t) + \sum_{i=1}^{n} b_i \chi_i(t) \qquad (8.41)$$

where the $\{a_i\}$ are the mixing coefficients associated with the state variable and the $\{b_i\}$ are the mixing coefficients associated with the adjoint variable. $\phi_0(t)$ satisfies the inlet condition on x; the $\{\phi_i(t)\}$ satisfy homogeneous inlet conditions. $\chi_0(t)$ satisfies the terminal adjoint variable boundary condition; $\{\chi_i\}$ satisfy the homogeneous condition at $t = t_N$.

The mixing coefficients are selected so that the approximations satisfy the system equations in an integral average sense. That is the residual of the state, and adjoint differential equations are required to be orthogonal to selected coordinate functions

$$\int_{t_0}^{t_N} \left[\frac{d\tilde{x}}{dt} - f(\tilde{x}, \tilde{V}) \right] \Psi_i dt = 0$$

$$i = 1, 2, \ldots, n \qquad (8.42)$$

$$\int_{t_0}^{t_N} \left[\frac{d\tilde{\lambda}}{dt} + \frac{\partial r(\tilde{x}, \tilde{V})}{\partial x} + \tilde{\lambda} \frac{\partial f}{\partial x}(\tilde{x}, \tilde{V}) \right] \Psi_i dt = 0$$

$$i = 1, 2, \ldots, n \qquad (8.43)$$

where the various functions are evaluated at $\tilde{x}$, $\tilde{\lambda}$ and $\tilde{V}$. The approximation, $\tilde{V}$, is obtained by requiring that the approximate Hamiltonian be maximized by $\tilde{V}$. The necessary condition for this is

$$\frac{\partial H(\tilde{x}, \tilde{\lambda}, \tilde{V})}{\partial V} = \frac{\partial r(\tilde{x}, \tilde{V})}{\partial V} + \tilde{\lambda} \frac{\partial f(\tilde{x}, \tilde{V})}{\partial V} = 0. \qquad (8.44)$$

Equation (8.44) provides an implicit relationship among $\widetilde{x}$, $\widetilde{\lambda}$ and $\widetilde{V}$. The implicit function theorem guarantees that an explicit representation for $\widetilde{V}$ can be obtained

$$\widetilde{V} = V\ (\widetilde{x},\ \widetilde{\lambda})$$

since the second partial of the Hamiltonian with respect to V must not vanish if the problem is to have a solution.

One advantage of this approach to solving the problem is that it does not require the solution of differential equations. We have seen that an approximation of the decision policy requires that the system differential equations be solved before parameter optimization can be carried out. In this case, such solution is avoided by simply requiring that the approximations satisfy the system equations in an integral average sense. This immediately produces the required set of algebraic equations in terms of the mixing coefficients.

A second advantage to the method is that it allows a knowledge of the general shape of the state variable trajectories to be used to suggest trial functions. This knowledge is often available in physical situations, and the method provides a formal mechanism for incorporating it into the solution to the problem.

Example 8.7 Resolve the problem of example 8.6 by using the trajectory approximation approach.

Solution The Hamiltonian is formed as

$$H = \lambda\ [V^2 - (V^2 + V)x\,],$$

and the minimizing condition is

$$\widetilde{V} = \frac{\widetilde{x}}{2(1 - \widetilde{x})}\ .$$

If this condition is substituted into the state variable equation, we obtain

$$\frac{d\widetilde{x}}{dt} = \frac{-\ \widetilde{x}^{\,2}}{4(1 - \widetilde{x})}\ .$$

Since the adjoint variable does not enter the equation, we need only to approximate x. A suitable form is

$$x \approx \widetilde{x} = 0.9 + at^{1/2}.$$

This approximation satisfies the inlet boundary condition and has the general shape expected of a reacting species.

The mixing coefficient, a, is selected to satisfy equation (i) in a weighted residual sense. By using the trial function as the weighing factor, we obtain

$$\int_0^1 \left[\frac{d\widetilde{x}}{dt} + \frac{\widetilde{x}^2}{4(1 - \widetilde{x})} \right] t^{1/2} \, dt = 0$$

which we can write as

$$\int_0^1 \left[\frac{d\widetilde{x}}{dt} - \frac{(1 + \widetilde{x})}{4} + \frac{1}{4(1 - \widetilde{x})} \right] t^{1/2} \, dt = 0.$$

Substitution for $\widetilde{x}$ gives

$$\int_0^1 \left[\frac{1}{2} \, at^{-1/2} - \frac{1}{4} (1.9 + at^{1/2}) \right. $$
$$\left. + \frac{1}{4(0.1 - at^{1/2})} \right] t^{1/2} \, dt = 0.$$

Each term in this equation can be integrated directly to obtain the following equation for a:

$$\frac{3a^4}{4} - \frac{3.8a^3}{6} - \frac{1}{2} a^2 - 0.1a - 0.01 \, \ell n \left(1 - \frac{a}{0.1} \right) = 0$$

which can be solved to yield

$$a^* = -0.38.$$

This gives for $\widetilde{V}$ the following:

$$\widetilde{V} = \frac{0.9 - 0.38t^{\frac{1}{2}}}{0.2 + 0.76t^{\frac{1}{2}}}.$$

A plot of $\widetilde{V}$ versus t is given in Figure 8.6 below. Also plotted is V^*. The agreement between $\widetilde{V}$ and V^* is quite good, indicating that the approximation technique is quite acceptable in this case. It should also be noted that a single parameter was used to obtain the approximate solution and that the equation which provided the numerical value for the parameter could be obtained in closed form.

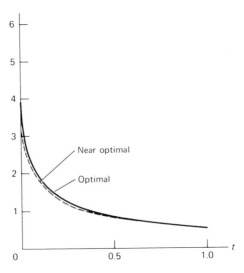

Figure 8.6

Example 8.7 presented a particularly simply trajectory optimization problem, one in which the optimal decision policy did not depend explicitly on the adjoint variable. Problems of this sort are called disjoint. The effectiveness of the trajectory approximation technique is by no means limited to disjoint problems, as the next example will show.

Example 8.8 Obtain an approximate solution to the problem of maximizing the yield of B at the outlet of a tubular reactor in which the following reaction scheme is taking place

$$A \underset{k_2}{\overset{k_1}{\underset{\rightarrow}{\leftarrow}}} B \overset{k_3}{\rightarrow} C.$$

In order to be specific, consider the case where

$$k_1^2 = k_2 = \frac{1}{2} k_3 = V^2$$

and

$$0 \leqslant V \leqslant 1.$$

The inlet conditions are

$$x_1(0) = 1, \quad x_2(0) = 0$$

where

$$x_1 = \text{mole fraction of } A$$
$$x_2 = \text{mole fraction of } B.$$

Solution We have seen that the system equations can be written in terms of t, the dimensionless length

$$\frac{dx_1}{dt} = -Vx_1 + V^2x_2 \qquad \text{(i)}$$

$$\frac{dx_2}{dt} = Vx_1 - 3V^2x_2. \tag{ii}$$

The Hamiltonian is

$$H = \lambda_1 (-Vx_1 + V^2x_2) + \lambda_2 (Vx_1 - 3V^2x_2) \tag{iii}$$

where

$$\frac{d\lambda_1}{dt} = (\lambda_1 - \lambda_2)V \tag{iv}$$

$$\frac{d\lambda_2}{dt} = -(\lambda_1 - 3\lambda_2)V^2 \tag{v}$$

and because we wish to maximize $x_2(1)$

$$\lambda_1(1) = 0, \quad \lambda_2(1) = 1. \tag{vi}$$

In terms of the approximations $\widetilde{x}_1$, $\widetilde{x}_2$, $\widetilde{\lambda}_1$ and $\widetilde{\lambda}_2$, the near-optimal policy, $\widetilde{V}$, is

$$\widetilde{V} = \frac{\widetilde{x}_1 (\widetilde{\lambda}_1 - \widetilde{\lambda}_2)}{2\widetilde{x}_2 (\widetilde{\lambda}_1 - 3\widetilde{\lambda}_2)} \quad \text{if this value} \leqslant 1$$

$$\widetilde{V} = 1 \quad \text{if above value} > 1.$$

Since $x_2 = 0$ at $t = 0$, the value called for by the unconstrained optimization of H at $t = 0$ would be infinite. Because of the restriction on V, however, we must limit the actual value to less than one; hence, the two conditions on $\widetilde{V}$ above.

In this case, the near optimal policy depends on both system variables and both adjoint variables. Hence, the trajectories for each of these variables must be approximated. Suitable forms are

$$\tilde{x}_1 = 1 + a_1 t^{1/2}$$

$$\tilde{x}_2 = a_2 t^{1/2}$$

$$\tilde{\lambda}_1 = b_1 (t - 1)$$

$$\tilde{\lambda}_2 = 1 + b_2 (t - 1).$$

The approximate control, $\tilde{V}$, is

$$\tilde{V} = \frac{1 + a_1 t^{1/2} \ [(b_1 - b_2)(t-1) \ - \ 1]}{2a_2 t^{1/2} \ [(b_1 - 3b_2)(t-1) \ - \ 3]} \quad \text{if this value} \ \leqslant \ 1 \quad \text{(vii)}$$

$$\tilde{V} = 1 \ \text{if above value} \ \geqslant \ 1.$$

The state variable equations are then solved approximately with this form for $\tilde{V}$ and the above approximations for x_1 and x_2 by the method of weighted residuals. Similarly the adjoint equations are solved approximately. This procedure generates four algebraic equations in four unknowns —a_1, a_2, b_1 and b_2. In this case and in general, the solution must be obtained numerically.

The values for a_1, a_2, b_1 and b_2 have been obtained by Zahradnik and Parkin (12) to be

$$a_1 = - \ 0.415 \qquad b_1 = 0.239$$

$$a_2 = 0.283 \qquad b_2 = - \ 0.649.$$

The near-optimal control trajectory obtained by substitution of these values into equation (vii) is shown below along with the solution obtained by solving the system and adjoint equations exactly by the approximation-to-the-problem method discussed in section 8.2.1. The accuracy of the approximation is quite good, and the computing time required to obtain the solution is minimal.

The results from this example show that trajectory approximation is an adequate alternative to the exact solution of the two-point boundary value problems which occur in trajectory optimization problems. In many cases, this alternative is quite desirable — particularly when the system model itself poses a two-point boundary value problem. In order to illustrate this situation, we consider maximizing the yield from a tubular reactor where axial dispersion is significant.

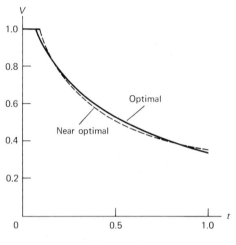

Figure 8.7

Example 8.9 Resolve the problem posed in example 8.8 if the effects of axial dispersion must be accounted for in the system equations.

Solution The system equations in this case are written as

$$\frac{1}{Pe} \frac{d^2x_1}{dt^2} - \frac{dx_1}{dt} = k_1x_1 - k_2x_2 \tag{i}$$

$$\frac{1}{Pe} \frac{d^2x_2}{dt^2} - \frac{dx_2}{dt} = - k_1x_1 + (k_2 + k_3)\, x_2 \tag{ii}$$

where Pe is the axial Péclet number.

The boundary conditions are

$$x_1(0+) - \frac{1}{Pe} \frac{dx_1}{dt}(0+) = x_1(0-) = 1$$

$$x_2(0+) - \frac{1}{Pe} \frac{dx_2}{dt}(0+) = x_2(0-) = 0$$

$$\frac{dx_1}{dt}(1) = 0$$

$$\frac{dx_2}{dt}(1) = 0.$$

The system equations are adapted to the form required by introducing two new state variables x_3 and x_4 as follows:

$$\frac{dx_1}{dt} = x_3$$

$$\frac{dx_2}{dt} = x_4.$$

Then, if the relationship among the rate constants assumed in the previous example is continued here, equations (i) and (ii) become

$$\frac{dx_3}{dt} = Pe \; (x_3 + Vx_1 - V^2x_2) \qquad \text{(iii)}$$

$$\frac{dx_4}{dt} = Pe \; (x_4 - Vx_1 + 3V^2x_2) \qquad \text{(iv)}$$

and the boundary conditions are

$$x_1(0) - \frac{1}{Pe} \, x_3(0) = 1$$

$$x_2 (0) - \frac{1}{Pe} x_4 (0) = 0$$

$$x_3 (1) = 0$$

$$x_4 (1) = 0.$$

It can be seen that the state variable equations themselves constitute a two point boundary value problem. This fact considerably complicates the determination of the optimal control policy and makes application of approximation-to-the-problem and approximation-to-the-solution methods extremely difficult. However, as we will show, trajectory approximation can be used to obtain an approximate solution to this problem with little more difficulty than encountered in the previous case.

First a Hamiltonian is formed:

$$H = \lambda_1 x_3 + \lambda_2 x_4 + \lambda_3 \ Pe(x_3 + Vx_1 - V^2 x_2)$$

$$+ \ \lambda_4 \ Pe(x_4 - Vx_1 + 3V^2 x_2)$$

where

$$\frac{d\lambda_1}{dt} = Pe V(\lambda_4 - \lambda_3)$$

$$\frac{d\lambda_2}{dt} = Pe V^2 (\lambda_3 - 3\lambda_4)$$

$$\frac{d\lambda_3}{dt} = - \lambda_1 - Pe\lambda_3$$

$$\frac{d\lambda_4}{dt} = - \lambda_2 - Pe\lambda_4$$

with the terminal boundary conditions obtained as before

$$\lambda_1(1) = 0$$

$$\lambda_2(1) = 1$$

and the initial conditions obtained by the more general transversality conditions as discussed in Chapter 7

$$\lambda_1(0) + Pe\lambda_3(0) = 0$$

$$\lambda_2(0) + Pe\lambda_4(0) = 0.$$

Because of the axial dispersion model, both the state and adjoint variable differential equations have boundary conditions specified at each end of the t-interval.

In order to apply trajectory approximation, we note that H is maximized by

$$V = \frac{x_1}{2x_2} \frac{(\lambda_3 - \lambda_4)}{(\lambda_3 - 3\lambda_4)}.$$

This explicit equation for V allows the near-optimal policy to be obtained upon solution of a set of algebraic equations. Suitable approximations for x_1 and x_2 which satisfy the stated boundary conditions are

$$\widetilde{x}_1 = 1 + a_1(t^2 - 2t - 2/Pe)$$

$$\widetilde{x}_2 = a_2(t^2 - 2t - 2/Pe).$$

Approximations for x_3 and x_4 which satisfy their differential equations exactly are

$$\widetilde{x}_3 = \frac{d\widetilde{x}_1}{dt} = 2a_1(t - 1)$$

$$\widetilde{x}_4 = \frac{d\widetilde{x}_2}{dt} = 2a_2(t - 1).$$

Simple, one parameter adjoint variable approximations for λ_1 and λ_2 are

$$\widetilde{\lambda}_1 = b_1(t - 1)$$

$$\widetilde{\lambda}_2 = 1 + b_2(t - 1).$$

If these expressions are substituted into the differential equations for λ_3 and λ_4 and the resultant equations integrated subject to the stated initial conditions, we can obtain approximations for λ_3 and λ_4 in terms of b_1 and b_2

$$\widetilde{\lambda}_3 = \frac{b_1}{Pe} \left(Pe + 1 - \frac{1}{Pe} e^{-Pet} - t \right)$$

$$\widetilde{\lambda}_4 = -\frac{1}{Pe} + \frac{b_2}{Pe} \left(Pe + 1 - \frac{1}{Pe} e^{-Pet} - t \right).$$

All the ingredients necessary to form the residuals for the system equations are now assembled. The number of mixing coefficients which must be obtained is four — similar to the case without axial dispersion.

The solution procedure involves setting up the residual equations for each of the variables x_1, x_2, λ_3 and λ_4. This results in four algebraic equations in four unknowns, whose solution provides the mixing coefficients which identify the near-optimal control policy. Values of these coefficients have been obtained for several Péclet numbers (13) and are reported in Table 8.5. The resultant control policies are shown on Figure 8.8 and compared to solutions obtained by other techniques.

Table 8.5. Table of Mixing Coefficients

	$Pe = 0.1$	$Pe = 1$	$Pe = 5$	$Pe = 10$
a_1	0.013693	0.103834	0.255851	0.314320
a_2	−0.009669	−0.072726	−0.177312	−0.217511
b_1	−0.201951	−0.207686	−0.210908	−0.207740
b_2	0.366659	0.431914	0.581186	0.656667

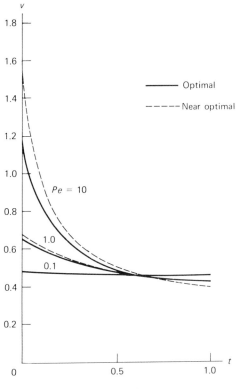

Figure 8.8

In addition to being able to generate good approximate solutions for trajectory optimization problems which involve models with boundary conditions specified at more than one location, trajectory approximation is effective in solving problems which are complicated by recycle streams or other nonserial structures. Such problems are often important in practice. Many engineering situations involve recycle systems and the ability to utilize a simple but effective scheme for their optimization is quite important.

The use of trajectory approximation to obtain near-optimal policies for recycle systems is illustrated in the following example.

Example 8.10 Let the reaction system of the previous two examples be carried out in a plug flow tubular reactor with a constant fraction of product recycle. The state equations in this case are

$$\frac{dx_1}{dt} = \frac{-Vx_1 + V^2x_2}{1 + r}$$

$$\frac{dx_2}{dt} = \frac{Vx_1 - 3V^2x_2}{1 + r}$$

where r is the recycle ratio.

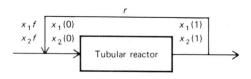

Figure 8.9

The boundary conditions on the state variables are

$$r\ x_1(1) + x_{1f} = (1 + r)\ x_1(o)$$

$$r\ x_2(1) + x_{2f} = (1 + r)\ x_2(o)$$

where x_{1f} and x_{2f} are the fresh feed mole fractions. It is once again desired to maximize the outlet concentration of B, $x_2(1)$.

Solution We form the Hamiltonian for this system:

$$H = \frac{\lambda_1(-Vx_1 + V^2x_2)}{1 + r} + \frac{\lambda_2(Vx_1 - 3V^2x_2)}{1 + r}$$

where

$$\frac{d\lambda_1}{dt} = \frac{V(\lambda_1 - \lambda_2)}{1 + r}$$

$$\frac{d\lambda_2}{dt} = - \frac{V^2(\lambda_1 - 3\lambda_2)}{1 + r}.$$

The boundary conditions in this case are obtained by the procedures discussed in section 7.5.1. They are

$$(1 + r)\, \lambda_1(1) - r\, \lambda_1(0) = 0$$

$$(1 + r)\, \lambda_2(1) - r\, \lambda_2(0) = (1 + r).$$

The optimal policy for the reactor is

$$V = \frac{x_1\,(\lambda_1 - \lambda_2)}{2x_2\,(\lambda_1 - 3\lambda_2)}.$$

Suitable trajectory approximations for the state and adjoint variables are

$$\widetilde{x}_1 = x_{1f} + a_1(t^{1/2} + r)$$

$$\widetilde{x}_2 = x_{2f} + a_2(t^{1/2} + r)$$

$$\widetilde{\lambda}_1 = b_1(t - 1 - r)$$

$$\widetilde{\lambda}_2 = (1 + r) + b_2(t - 1 - r).$$

The problem reduces to the determination of four mixing coefficients. These are obtained by forming the residuals for the four state and adjoint differential equations and setting their weighted residuals to zero.

Table 8.6 gives the mixing coefficients for three values of r. The resulting control policies are plotted on Figure 8.10 below.

Table 8.6

$r =$	0.1	1.0	10
a_1	−0.364	−0.169	−0.027
a_2	0.249	0.117	0.019
b_1	−0.236	−0.232	−0.207
b_2	0.626	0.451	0.371

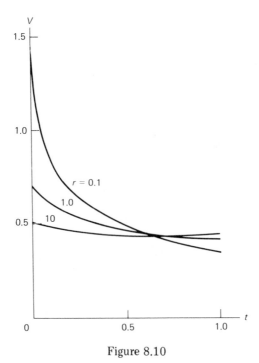

Figure 8.10

Example 8.11 Use trajectory approximation to obtain a near-optimal feedback law for the problem involving transversality conditions introduced in Example 7.11. Here $t_0 = 0$ and $t_N = 1$. The system equations are

$$\frac{dx_1}{dt} = x_2 \qquad (i)$$

$$\frac{dx_2}{dt} = V \qquad \text{(ii)}$$

subject to

$$x_1(o) = 0$$
$$x_2(o) = 0 \qquad \text{(iii)}$$

and

$$[x_1(1)]^2 + [x_2(1)]^2 = 1. \qquad \text{(iv)}$$

The adjoint equations are

$$\frac{d\lambda_1}{dt} = -x_1 \qquad \text{(v)}$$

$$\frac{d\lambda_2}{dt} = -x_2 - \lambda_1. \qquad \text{(vi)}$$

The transversality condition is

$$\lambda_1(1) \, x_2(1) - \lambda_2(1) \, x_1(1) = 0 \qquad \text{(vii)}$$

and

$$V^* = -\lambda_2. \qquad \text{(viii)}$$

If we approximate x_1 as

$$x_1 = a_1 t^2 + a_2 t^3.$$

Then, by equation (i),

$$x_2 = 2a_1 t + 3a_2 t^2.$$

Both approximations satisfy (iii). In order to satisfy (iv), we require that

$$(a_1 + a_2)^2 + (2a_1 + 3a_2)^2 = 1. \tag{I}$$

The approximation for λ_1 comes from integrating equation (v)

$$\lambda_1 = \frac{a_1}{3} t^3 - \frac{a_2 t^4}{4} + b_1$$

where b_1, the constant of integration, becomes an adjoint mixing coefficient. Substitution of the approximations for x_2 and λ_1 into equation (vi) and integration gives for λ_2

$$\lambda_2 = - a_1 t^2 - a_2 t^3 + \frac{a_1 t^4}{12} + \frac{a_2 t^5}{20} - b_1 t + b_2$$

where b_2 is another mixing coefficient.

The transversality condition is now approximately satisfied by

$$\left(b_1 - \frac{a_1}{3} - \frac{a_2}{4} \right) \left(2a_1 + 3a_2 \right)$$

$$= \left(b_2 - b_1 - \frac{11a_1}{12} - \frac{19a_2}{20} \right) \left(a_1 + a_2 \right). \tag{II}$$

Since $V^* = - \lambda_2$, we can now satisfy equation (ii) in a weighted residual sense:

$$\int_0^1 \left(2a_1 + 6a_2 t - a_1 t^2 - a_2 t^3 + \frac{a_1 t^4}{12} + \frac{a_2 t^5}{20} \right.$$

$$\left. - b_1 t + b_2 \right) \Psi_i(t) dt = 0. \quad i = 1, 2$$

We take $\Psi_1 = t$ and $\Psi_2 = t^2$ to obtain two algebraic equations

$$\frac{55a_1}{72} + \frac{253a_2}{140} - \frac{b_1}{3} + \frac{b_2}{2} = 0 \qquad \text{(III)}$$

$$\frac{67a_1}{140} + \frac{643}{480} a_2 - \frac{b_1}{4} + \frac{b_2}{3} = 0. \qquad \text{(IV)}$$

Equations (I-IV) now represent 4 algebraic equations in four unknowns, a_1, a_2, b_1, b_2. Their solution can be obtained by standard means. In particular, since equations (III) and (IV) are linear, the adjoint mixing coefficients can be easily expressed in terms of the state variable mixing coefficients by application of Cramer's Rule. Then equations (I) and (II) are expressed solely in terms of a_1 and a_2, and a solution can be obtained.

The mixing coefficients are

$$a_1 = 0.42 \quad b_1 = -0.35$$

$$a_2 = 0.02 \quad b_2 = -0.97.$$

The approximate trajectories obtained with these values are compared to the analytical solution from example 7.11 in Table 8.7 below.

Table 8.7. Exact and Approximate Solutions
for Example Problem

t	x_1^*	$\widetilde{x}_1$	x_2^*	$\widetilde{x}_2$
0.0	0.00	0.00	0.00	0.00
0.2	0.017	0.017	0.18	0.17
0.4	0.071	0.068	0.36	0.35
0.6	0.156	0.155	0.53	0.53
0.8	0.284	0.278	0.72	0.71
1.0	0.440	0.440	0.90	0.90

The near-optimal control policy is shown on Figure 8.11, along with the optimal policy from problem 7.11. The agreement here is not as good as with the state variables, and additional terms in the expansion of the adjoint variable expansions would be needed if greater accuracy is required.

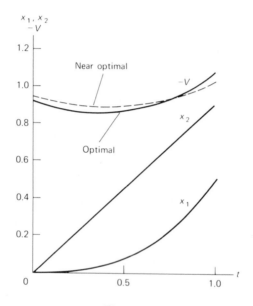

Figure 8.11

With the near-optimal control policy explicitly given, the approximations generated here can be used to develop a near-optimal negative feedback law for this problem. Thus we suggest

$$\widetilde{V} + b_2 = - k_1(t) \; x_1(t) - k_2(t) \; x_2(t)$$

where k_1 and k_2 are time varying feedback gain coefficients. In terms of our present approximation, let

$$k_1(t) = j_1 + j_2 t + j_3 t^2$$

$$k_2(t) = j_4 + j_5 t$$

where the j's are constants to be obtained. Other forms for k_1 and k_2 are possible; the proposed expansions are not unique. Moreover, they are only near optimal with respect to the present problem.

By substituting for $\widetilde{V}$, $\widetilde{x}_1$ and $\widetilde{x}_2$ into the near-optimal feedback formula, we obtain

$$- (j_1 + j_2 t + j_3 t^2)(a_1 t^2 + a_2 t^3) - (j_4 + j_5 t)(2a_1 t + 3a_2 t^2)$$

$$= b_1 t + a_1\ t^2 + a_2 t^3 - \frac{a_1}{12}\ t^4 - \frac{a_2}{20}\ t^5.$$

The j's can be obtained by equating the like powers of t on each side of the equation. This procedure yields

$$j_1 = \left(\frac{a_1^2}{15 a_2} + \frac{9}{2}\ \frac{a_2 b_1}{a_1^2} - 1 \right) = -0.6$$

$$j_2 = \frac{a_1}{30} = 0.014$$

$$j_3 = \frac{1}{20} = 0.05$$

$$j_4 = \frac{-b_1}{2a_1} = 0.426$$

$$j_5 = \left(\frac{a_1^2}{30 a_2} - \frac{3 b_1 a_2}{2 a_1^2} \right) = 0.357$$

so that the time varying near-optimal feedback gains are

$$k_1 = -0.6 + 0.014t + 0.05t^2$$

$$k_2 = 0.426 + 0.357t.$$

Thus we can envision the near-optimal feedback control scheme as illustrated in Figure 8.12. The scheme is only near-optimal for the problem and conditions posed and should not be continued beyond $t \geqslant 1$.

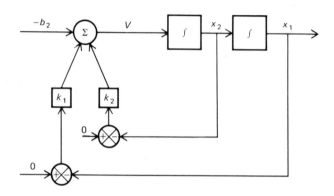

Figure 8.12

8.4 APPENDIX: NUMERICAL SOLUTION OF ORDINARY DIFFERENTIAL EQUATIONS

A first-order ordinary differential equation (ODE) and its initial condition can be written as

$$\frac{dy}{dx} = y' = f(x, y) \qquad (8.45)$$

$$y = y_0, \ x = x_0$$

To solve equation (8.45) numerically, we start with the known value of y_0 at x_0 and proceed to calculate y_1 at x_1, y_2 at x_2, etc. This is continued as far as necessary. There are two types of numerical procedures: (1) the *single-step method* is the one with which the value of y_{n+1} is calculated by using information at only a single other point y_n, and (2) the *multiple-step method* is the one which requires the information at several points in order to compute the new point y_{n+1}.

8.4.1 A Simple Technique — Euler's Method The Taylor series expansion of a function $y(x)$ about x_0 is

$$y(x) = y(x_0) + (x - x_0) y'(x_0)$$

$$+ \frac{(x - x_0)^2}{2!} y''(x_0) \ldots \tag{8.46}$$

If we truncate the series after the second term,

$$y(x) = y(x_0) + hy'(x_0), \quad h = x - x_0. \tag{8.47}$$

In doing so, we have implied that the terms beyond $y'(x_0)$ in (8.46) are small enough to be neglected. Therefore, to represent $y(x)$ by (8.47), an error (called truncation error) has been committed.

After combining (8.45) with (8.47), we have

$$y(x) = y(x_0) + hf(x_0, y_0) \tag{8.48}$$

and, since x and x_0 are completely arbitrary, (8.48) can be written as

$$y_{n+1} = y_n + hf(x_n, y_n). \tag{8.49}$$

This is the Euler method, which is the simplest numerical scheme to solve a first-order ODE. We notice that it is a single-step method.

If $f(x, y)$ is reasonably smooth or if y possesses a reasonably smooth derivation, f should not vary much between two successive values of x_n and x_{n+1}, and the magnitude of the terms omitted therefore depend on h. If h is small, the powers of h would be even smaller. It is sufficient to say the largest omitted term is proportional to h^2. The Euler's method is said to be of the order of h.

8.4.2 The Runge-Kutta Method and Gill's Modification It is fair to say, at present, that the most popular method for solving the initial value problem is the Runge-Kutta method. Although this is a single-step method, the associated truncation error is much less than that of the method previously outlined. We shall omit the derivation of this equation and present only the working formulas:

(i) third-order formula — accuracy: $O(h^4)$

$$k_1 = hf(x_n, \ y_n)$$

$$k_2 = hf\left(x_n + \frac{h}{2}, \ y_n + \frac{k_1}{2}\right)$$

$$k_3 = hf(x_n + h, \ y_n + 2k_2 - k_1)$$

$$y_{n+1} = y_n + \frac{1}{6}(k_1 + 4k_2 + k_3)$$

(ii) fourth-order formula — accuracy: $O(h^5)$

$$k_1 = hf(x_n, \ y_n)$$

$$k_2 = hf\left(x_n + \frac{h}{2}, \ y_n + \frac{k_1}{2}\right)$$

$$k_3 = hf\left(x_n + \frac{h}{2}, \ y_n + \frac{k_2}{2}\right)$$

$$k_4 = hf(x_n + h, \ y_n + k_3)$$

$$k_{n+1} = y_n + \frac{1}{6}(k_1 + 2k_2 + 2k_3 + k_4)$$

(iii) Gill's modified fourth-order formula

$$k_1 = hf(x_n, \ y_n)$$

$$k_2 = hf\left(x_n + \frac{h}{2}, \ y_n + \frac{k_1}{2}\right)$$

$$k_3 = hf \left[x + \frac{h}{2}, \ y_n + \left(-\frac{1}{2} + \frac{1}{\sqrt{2}} \right) k_1 \right.$$

$$\left. + \left(1 - \frac{1}{\sqrt{2}} \right) k_2 \right]$$

$$k_4 = \left[hf \ x_n + h, \ y_n + \left(-\frac{1}{\sqrt{2}} \right) k_2 + \left(1 + \frac{1}{\sqrt{2}} \right) k_3 \right]$$

$$y_{n+1} = y_n + \frac{1}{6} \left(k_1 + k_4 \right) + \frac{1}{3} \left(1 - \frac{1}{\sqrt{2}} \right) k_2$$

$$+ \frac{1}{3} \left(1 + \frac{1}{\sqrt{2}} \right) k_3$$

As can be seen from the above, the amount of computation required in the Runge-Kutta method is much more than that of the simpler Euler method outlined above. However, with the aid of a computer, this poses no particular problem. Various versions of the Runge-Kutta method have been programmed in subroutines, are available in most computer systems, and can be called easily.

Bibliography

8.1 INTRODUCTION

The classification of methods used in this section has been suggested by several authors. Further discussion on the subject along with many applications of variational methods to control problem is found in

1. Lapidus, Leon and Rein Luus, *Optimal Control of Engineering Processes*, Blaisdell Publishing Co., Waltham, Massachusetts, 1967.

8.2 INDIRECT METHODS

Texts dealing with solution techniques include

2. Bryson, Arthur E. and Yu-Chi Ho, *Applied Optimal Control*, Blaisdell Publishing Co., Waltham, Massachusetts, 1969.
3. Denn, Morton M. *Optimization by Variational Methods*, McGraw-Hill Book Co., New York, 1969.
4. Luenberger, David G. *Optimization by Vector Space Methods*, John Wiley & Sons, New York, 1969.

Many applications of variational calculus to engineering problems have appeared. The following are compilations of some of these:

5. Athans, Michael and Peter L. Falb, *Optimal Control*, McGraw-Hill Book Co., New York, 1966.
6. Balakrishnan, A.V. and L.W. Neustadt (eds), *Computing Methods in Optimization Problems*, Academic Press, Inc., New York, 1964.
7. Balakrishnan, A.V. and L.W. Neustadt (eds), *Mathematical Theory of Control*, Academic Press, Inc., New York, 1962.
8. Leitmann, G. (ed), *Optimization Techniques*, Academic Press, Inc., New York, 1962.

9. Leitmann, G. (ed), *Topics in Optimization*, Academic Press, Inc., New York, 1967.

8.3 DIRECT METHODS

The approximate solution of differential equations by the method of weighted residuals is discussed in

10. Collatz, L. *The Numerical Treatment of Differential Equations*, Springer-Verlag, Berlin, 1960.
11. Rice, John R. *The Approximation of Functions; vol I: Linear Theory*, Addison-Wesley Publishing Co., Inc., Reading, Massachusetts, 1964.

The use of approximation theory to solve trajectory optimization problems is described in the following papers:

12. *Zahradnik, Raymond* L. and Elliot S. Parkin, Computation of near-optimal temperature profiles for a tubular reactor in *Computing Methods in Optimization*, Academic Press, Inc., pp. 389-397, 1969.
13. Lynn, LeRoy L., Elliot S. Parkin and Raymond L. Zahradnik, Near-optimal Control by Trajectory Approximation. Tubular Reactors with Axial Dispersion, *Ind. Eng. Chem. Fundam.*, 9, pp. 58-62, 1970.
14. Zahradnik, Raymond L., Elliot S. Parkin and LeRoy L. Lynn, *Approximate Models for Optimal Control of Chemical Processes*, Summer Computer Simulation Conference, Denver, Colorado, pp. 301-308, 1970.

Index